Biochemistry Lab Manual

David A. Thompson

Biochemistry Lab Manual
David A. Thompson
Publication date 2018
Copyright © 2018 David A. Thompson

An instructor's manual can be obtained by contacting the author at thomp@post.harvard.edu.

Prior warning is not given for every chemical substance a student or instructor may be exposed to while engaged in the activities described in this text. It is the responsibility of the individual supervising any laboratory exercise to ensure that those individuals engaging in the activity are made aware in advance of any risks or hazards associated with that activity. Hazards associated with a chemical exposure should be described in the Safety Data Sheet (SDS) for the compound as provided by the manufacturer or distributor.

The course instructor and/or the individual supervising the laboratory exercises share the ultimate responsibility for the welfare of the student in the laboratory, ensuring that actions in the laboratory are performed in a safe and prudent fashion. These responsibilities include (1) exercising good judgement in planning, conducting, and supervising any experimental exercise, (2) maintaining a safe laboratory environment (this includes both maintaining laboratory equipment in safe condition and ensuring the availability of appropriate safety equipment), and (3) documenting that students have received thorough safety instruction.

To encourage a safe lab experience, an instructor should (1) consider demonstrating any component of an activity which may involve a risk to the student, (2) maintain continuous supervision of students involved in the activity, and (3) ensure he/she has sufficient training to handle any reasonably foreseeable emergency.

Contents

Contents

Contents

1. Laboratory reports

1.1. Data collection

Before preparing a written scientific report, one typically has accumulated some data of interest and carefully recorded that data. Scientific data is recorded and stored in many different ways. If the records are likely to ever be associated with any legal proceeding, they should be kept meticulously in a bound notebook or using a computerized record-keeping system. In such circumstances, after entering the data, the record can be verified with the signature and date of a witness or by using the verification functionality of a computer-based record-keeping system.

In an undergraduate lab, this type of record-keeping is over-kill. Nevertheless, one should keep meticulous records. Your instructor may have specific expectations with respect to how you keep records. Such expectations might include the use of only ink in recording data, a specific manner in which any errors in recording are indicated and corrected, or the use of a specific medium for recordkeeping (e.g., a bound notebook).

1.2. Guidelines for preparing lab reports

1.2.1. Format guidelines

The report should be

- in blue or black ink
- either clearly and neatly handwritten, typed, or printed from a computer
- left-justified, double-spaced, with one-inch margins on the left and right sides
- if printed from a computer, in a font which is normal weight 12 pt serif
- formatted for printing on 8.5 x 11 inch paper
- composed of numbered pages

1.2.2. Organization of the report

The report should be arranged with the following sections:

1. title, date of submission, author(s)
2. introduction
3. methods (experimental procedures)
4. results
5. discussion
6. references
7. figure legends
8. tables
9. figures
10. supplemental data

1.2.3. The sections of the report

Some of the texts listed in the 'selected references' section below provide excellent guidelines for writing a scientific paper. Familiarize yourself with these guidelines. Another approach to better understand the organization of a scientific paper and the content of each section is to read scientific articles.

Each section of the report should be clearly labeled.

Prose should be concise and unambiguous, avoiding a casual, conversational style.[1] Typically, this includes avoiding use of the first person.[2]

Typically, a scientific publication/report also includes an **abstract**. The abstract is a short summary of the document. It should give a good overview of the results and be **very** accurate. The abstract is what will be read by most individuals; few individuals will read the complete paper.

The **title section** should include the title (as short and informative as possible), the date of submission, and a list of authors (names of all authors and corresponding email addresses). Ensure that you exercise care in listing authors. For example, if the report describes a collaborative project, it may be appropriate to list several individuals as authors. However, it is likely that your instructor will expect you to clearly specify who actually wrote the text of the report.

The **introductory section** describes the subject of the study and the rationale/motivation underlying the study. Previous experiments leading to the current study are summarized (and the appropriate papers cited) in such a fashion that the reader is able to see how the train of logic progresses through the previous experiments to the question experimentally addressed in the present paper. Frequently this question is presented as an explicit hypothesis. In this course, you are not expected to complete a thorough literature search in order to write the introduction (unless specif-

---

[1] An example of poor style: "Well, we conducted a study in our microbiology lab and started off by examining two restroom facilities..."

[2] For example, typically "...40 mL were heated..." is preferable to "...we heated 40 mL...".

ically indicated); instead, a simple summary of the hypothesis of the study, the methodology used to test the hypothesis, and the conclusion drawn from the study will suffice.

The **methods section** describes the methodology used in the study. This section should be written in prose in the past tense and should be written in a sufficiently detailed fashion that the experiment could be reproduced by the reader.[3]

The **results section** describes the data obtained (the observations) from the present study. Typically, results are also presented in figures and/or tables. This section should include prose describing the data represented in the figures or tables as well as any other pertinent results. One should be able to, by reading the prose of the results section, know what each figure communicates. The text should contain actual numerical values (e.g., fold-elevation, % apoptotic, ...). The prose should explicitly refer to figures and tables as appropriate.[4] Often, the results described in the "Results" section of a paper, are followed by one or several brief conclusions drawn from the results of each experiment. If a theoretical description (such as a mathematical equation) of the experimental system being studied is presented in the lecture or the lab handout, the results section is typically the appropriate place to apply it to your data. Occasional errors include omission of titles for figures or tables, omission of numbers for figures or tables, omission of figure legends, and omission of the prose component of the results section.

The **discussion section** summarizes the data and then proceeds to the author's interpretation of the data. Does the data support the hypotheses presented in the introduction (or elsewhere...)? How will this study impact future stud-

---

[3]Frequent errors include (1) composing this section as a list of numbered steps rather than writing it in prose and (2) using the imperative rather than the past tense.

[4]An example: "Two bands of approximately 98 and 110 kDa were observed in lane 2 of the gel (Fig. 2)"

ies?[5] Does the study have larger social or societal ramifications? Does the data allow the author to construct a model of how a particular biological process occurs?

The **supplemental data** section (optional) may contain a copy of all raw data as well as any detailed calculations and formulas used during analysis of experimental data.

1.2.4. Tables and Figures

Each table or figure should

- be labeled with 'Table' or 'Figure' (if it's not a table, it's a figure...)

- be titled (e.g., "Table 2. Position versus velocity")

- be numbered (e.g., "Figure 3. Dependence of absorption on time")

- be accompanied by a descriptive legend

Figures should have descriptive axis labels (including units of measure, when appropriate). The axis should be informative: "Frequency / 10000" is worse; "Frequency of premature chromatin condensation / 10000 cells" is better.

1.3. Presenting data

Whether one continues in science or enters another field, it is important to understand how to interpret and present data in the context of "graphs" or "charts". Interpreting raw data typically requires a substantial investment of effort. One of the author's "jobs" is to reduce the amount of effort the reader must invest in order to visualize data. When

---

[5]In other words, what questions do the data suggest should be asked next? How would those questions be addressed experimentally in future studies?

week	mass (g)
8	0.03 0.05 0.07 0.09 0.06
12	0.03 0.05 0.07 0.08 0.12 0.19
16	0.05 0.07 0.07 0.09 0.15
20	0.05 0.07 0.08 0.1 0.2

Table 1.1. Thymus masses at different mouse ages

done well, the presentation enables the reader to comprehend trends in data and aids the reader in evaluating the conclusions drawn with respect to the hypothesis/es presented by the author.

For example, consider a (hypothetical) study of thymus mass versus mouse age. While thymus mass tends to rise with age, this trend isn't immediately apparent from inspection of the table of raw data (Table 1.1). Neither does a scatterplot of the data (Figure 1.1) forcefully drive this point home. If it is desirable to emphasize the trend, the point might be driven home more forcefully by using a figure with statistical summaries (Figure 1.2).

1.3.1. References

When material from another source is used in the report, the source must be acknowledged. This is done through (1) an in-text citation and (2) the inclusion of a reference list (typically near the end of the document). Use the APA citation and reference list styles (see http://www.apastyle.org/) when preparing a report.

Acceptable references and sources include articles in peer-reviewed journals or peer-reviewed texts. The primary reference should be used whenever possible. Other references and sources are generally not acceptable. For example, textbooks are generally not acceptable references. Neither are many "web pages" (e.g., wikipedia articles) acceptable ref-

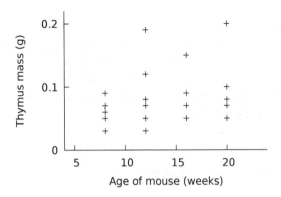

Figure 1.1. Scatterplot of thymus mass data. Mice were sacrificed at the indicated ages. The wet weight of the thymus of each sacrificed mouse (solid cross) was determined immediately after euthanization of the mouse.

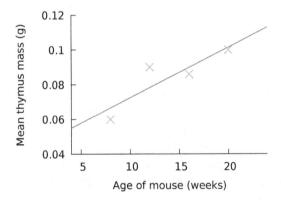

Figure 1.2. Statistical summaries of thymus mass data. Mice were sacrificed at the indicated ages. The wet weight of the thymus of each sacrificed mouse was determined immediately after euthanization of the mouse. A least-squares linear regression was performed (solid line) using averages at each time point.

erences.

If an excerpt from another work is used verbatim, it must be clearly formatted as such. Short quotes can be indicated by surrounding the quoted material with double-quote characters. Longer quotes should be formatted as indented block-quotes. Both short and long quotes should be accompanied by an appropriate citation.

1.4. Additional resources

See Appendix B for some suggestions on improving one's writing. For additional guidance in your writing, consider consulting one of the many texts on scientific writing. [1, 19, 6, 8, 14, 15, 17]

Several high-quality word-processing packages are freely available including LibreOffice (`http://www.libreoffice.org/`), OpenOffice (`http://www.openoffice.org`), and Abi-Word (`http://www.abisource.com`).

A number of packages are freely available for generating charts, graphs, and plots. The gnumeric (`http://projects.gnome.org/gnumeric/`), LibreOffice (`http://www.libreoffice.org/`), and OpenOffice (`http://www.openoffice.org`) spread-sheet components all have some charting capabilities. The stand-alone charting program gnuplot (`http://www.gnuplot.info`) is highly configurable, capable of two- and three-dimensional plotting, curve fitting (using an arbitrary function), and generating histograms.

2. Safety

Before beginning any work in the laboratory, the course in-structor should discuss the laboratory rules[1] and any spe-cific safety issues associated with the laboratory component of the course. Some lab procedures may have specific safety considerations; these will be noted in the laboratory manual and/or described by the course instructor prior to the lab.

---

[1]Appendix A contains an example of some guidelines which might be enforced in the laboratory component of a university course.

3. Paper wads

3.1. Introduction

To the uninitiated, paper wads and gravity may seem to have little relationship to biochemistry.

3.2. Procedure

This experiment is a "group experiment". All that will be needed is a set of paper wads (four per group member).

3.2.1. Experiment 1

1. Each class member should acquire or prepare four paper wads (crush a US letter-sized paper sheet into a crumpled ball).

2. The instructor will divide the class into two groups, "A" and "B", of approximately equal size.

3. The two groups should position themselves across from each other. The instructor will indicate where members of "Group A" and "Group B" should position themselves.

4. All members of "Group A" will begin with all the paper wads. Members of "Group B" will begin empty-handed.

5. Variables of interest include the number of paper wads Group A possesses, the number of paper wads Group B possesses, and the time.

6. Record the data associated with the first time point ($t = 0s$).

7. When the instructor indicates, begin throwing paper wads at the members of the other group. Rules:

   - **You may only throw one paper wad at a time.**

   - **You must stay on "your side" of the line dividing the two groups.**

8. When the instructor indicates, stop and record data.

3.2.2. Experiment 2

1. Members of "Group A" will give "Group B" members all their paper wads (i.e., the exercise begins with *all* of the paper wads in possession of "Group B"). Members of "Group B" will head upstairs to a position opposed to "Group A". The instructor will indicate where members of "Group A" and "Group B" should position themselves.

2. The variables of interest are the same as in the previous experiment.

3. Record the data associated with the first time point

4. When the instructor indicates, begin throwing paper wads at the members of the other group.

5. When the instructor indicates, stop and record data.

3.3. Laboratory report

For this report,

1. Focus on preparing an excellent "results" section and an excellent "discussion" section.

2. In addition, answer the following questions and include your answers on a separate (from the lab report) sheet of paper.

The equilibrium constant for the reaction $A \rightleftharpoons C$ is 1. Assume a very low energy of activation for the reaction.

- You prepare a 50 mM solution of A and return to analyze the solution after a week has passed. Assuming sterile conditions, predict the concentration of A and the concentration of C in the solution after one week.

- At this point, are molecules of A still converting to C? Are molecules of C still converting to A?

The equilibrium constant for the reaction $B \rightleftharpoons D$ is 2. Assume a very low energy of activation for the reaction.

- You prepare a 50 mM solution of B and return to analyze the solution after a week has passed. Assuming sterile conditions and the absence of significant side-reactions,[1] predict the concentration of B and the concentration of D in the solution after one week.

- At this point, are molecules of B still converting to D? Are molecules of B still converting to D?

Is the statement "the system is at equilibrium" synonymous with the statement "chemical transformations are no longer occurring in the system"?

Is the statement "the system is at equilibrium" synonymous with the statement "the chemical components of the system are all present at equal concentrations"?

---

[1]i.e., under these conditions, the only reaction B tends to undergo is conversion to D and the only reaction D tends to undergo is conversion to B

3.4. Acknowledgements

Orvis and Orvis provided a useful starting point for developing this lab. [22]

4. Buffer preparation

4.1. Prior to the day of the lab

Review the definitions of K_a, pH, molarity (the sum of all species of the acid), and osmolarity.

1. Select a buffer from the list below:

   phosphate buffer (8.5 mM phosphate, pH 7.4)[1]
   acetate buffer (24 mM acetate, pH 5.0)

2. Be prepared to hand in, *at the start* of the laboratory session, a copy of the calculations described below:

   - Determine the pK_a value(s) of the buffer you selected. Use that data to calculate the concentrations and masses of the acid and base forms present in 100 mL of the buffer solution. You may find a reference such as the CRC or the Merck Index a useful source of K_a and/or pK_a values. You may find the *Henderson-Hasselbalch equation* useful in your calculations.

   - Prepare a recipe for preparing the buffer (see "Available reagents", below, when preparing your recipe).

3. At the beginning of class, write your name in the column succeeding the buffer name on the board.

---

[1]When a buffer is referred to as "10 mM phosphate" or "40 mM oxalate", the number is typically intended to refer to the **total** molarity. For example, a "10 mM phosphate buffer" consists of H_3PO_4, $H_2PO_4^-$, HPO_4^{2-}, and PO_4^{3-} where the sum of the molarities of all four species is 10 mM.

4.2. Introduction

In this lab, you will prepare a buffer solution of a given pH.

The ability to prepare a solution of defined pH and composition is essential in most fields of biology and chemistry. In biochemistry, for example, an enzyme may have an activity at pH 5 that is dramatically different than the activity of the enzyme at pH 8. For an assay for the enzyme (e.g., a clinical assay of enzyme levels in blood) to yield consistent data, the solution in which the assay is conducted must have a near-constant pH. A second example: in clinical situations requiring intravenous infusions of a fluid, it is essential the the fluid be of an osmolality very close to that of blood. Even small deviations in blood osmolality can cause convulsions, cardiac arrest, and/or death.

4.3. Procedure

1. Prepare 100 mL of the buffer solution using the available reagents. Since adding solids and liquids to a 100 mL volume will increase the volume, it is best to:

   - dissolve all components in a volume 80-90% of the desired final volume

   - add a stir bar and mix to dissolve all components

   - adjust the volume to the final volume in a graduated cylinder or volumetric flask only after all components are dissolved and the stir bar is removed

2. Review the guidelines for using a pH meter (see Appendix E).

3. Measure the pH of your buffer using a pH meter.

4.3.1. available reagents

- acetic acid, 17.4N (glacial)

- sodium acetate

- hydrochloric acid, 2M

- sodium chloride

- phosphoric acid, 14.7N (85%)

- sodium phosphate, monobasic, monohydrate (NaH_2PO_4 x H_2O)

- sodium phosphate, dibasic (Na_2HPO_4), anhydrous

- sodium phosphate, tribasic, dihydrate (Na_3PO_4 x $2H_2O$)

4.4. Laboratory report

Items you may wish to consider:

1. Ensure you describe methods used to assess each volume and mass

2. Ensure you consider both the hypothetical and measured pH of the solution

Ensure you include, as an appendix, the calculations you used in preparing the buffer.

On a separate sheet, consider an additional question: if you were going to prepare the solution you chose, but as a 150 mOsm solution, describe the "recipe" you would use if you were using NaCl to adjust osmolarity.

4.5. Acknowledgements

Parts of this lab were originally adapted from a similar exercise described by V. Vouillet[2].

---

[2]Originally described by V. Vouillet (Smith College) in 2001 online; however, the exercise is no longer accessible online.

5. Pipet calibration

5.1. Introduction

In conducting an experiment, the quality of data obtained depends on the quality of the measurement. The quality of a measurement is related both to the technical skill of the experimenter and to the quality and calibration of the instrument he or she is employing to make the measurement. The quality of the measurement is characterized by both the accuracy and the precision of the measuring system. Ensure you have a clear sense of the definition of each term.

In practice, with experimental measurements, it is important not only to make a measurement that is valid but to be able to present one's data in a clear and meaningful fashion. In both academia and business, such presentations are regularly made to one's peers and to one's employer or funding source(s). It becomes important not only to have quality data, but to have the ability to convey the message behind the data in a convincing and clear fashion. Tufte and Tukey have both emphasized the need to exercise care in developing a visual summary or representation of data. The recent trend toward analysis of larger datasets, networks, and multivariate datasets brings additional challenges in visualization of data [12].

To test hypotheses regarding data, statistical methods are frequently used. One common statistical expression is the **confidence interval**. A confidence interval on the mean of a population is an estimate of the actual, true mean of the population - it is an interval which "traps" the mean with some degree of certainty. For example, the 95% confidence interval (L1,L2) suggests that, for a given set of 100 measurements, we expect 95 the measurements to fall between L1 and L2. If the data are assumed to follow a normal distribution then a 95% confidence interval can be expressed as

$$\left(x - \frac{1.96}{\sqrt{n}}, x + \frac{1.96}{\sqrt{n}}\right)$$

where

x = measured mean
σ = standard deviation
n = number of samples

5.2. Procedure

1. Obtain a pipettor with a range which includes 1000 µL (some may have slightly different ranges). Obtain tips for the pipettor.

2. Record the identity of the pipettor.

3. Tare a piece of weigh paper or a weigh boat and add 1000 µL water to the paper; record the measurement.

4. Tare the same piece of paper, add an additional 1000 µL and record the measurement.

5. Repeat the process a third time.

6. Obtain a pipettor with a range which includes 2 µL (some may have slightly different ranges). Obtain tips for the pipettor.

7. Tare a piece of weigh paper or a weigh boat and add 2 µL of DMSO to the paper; record the measurement.

8. Repeat the measurement twice (in the same fashion as you did for the 1000 µL measurements).

5.3. Laboratory report

The report should be a fairly short report following the format described in the first lab session.

Before beginning to write the report in earnest, stop and think about a useful approach to scientific analysis: for each datum obtained, describe the corresponding 'expected value' (e.g., how much do you *expect* 1 mL of water to weigh?).

Then formulate a sensical and simple hypothesis. As you prepare the report, ensure you do not neglect to explicitly state your hypothesis(es). Finally, contrast the 'observed values' with the 'expected values' using statistical tools and draw conclusion(s) regarding the hypotheses (i.e., do the data support hypothesis X?).

Ensure you use a 95% confidence interval to describe each data set. Use the calculated intervals to draw conclusions with respect to the hypothesis(es?) of interest.

Finally, ensure you reference any values used in the paper which are not generally known or accepted (e.g., the density of DMSO).

5.4. Additional resources

Introductory statistics texts typically contain tables describing the cumulative standard normal and cumulative binomial distributions.

Modern calculators that can execute common statistical functions can be purchased relatively inexpensively.

A number of computer software packages can be used to calculate various statistics (see Appendix D for examples).

The Merck Index, the CRC Handbook, and a variety of chemical catalogs (e.g., the Sigma-Aldrich catalog and the ICN catalog) contain physical constants (e.g., density, melting point, boiling point) for many compounds.

6. Identifying a buffer by titration

6.1. Introduction

The typical introductory undergraduate chemistry course introduces the concepts of acids and bases. Some molecules contain multiple functional groups able to accept and release protons. An amino acid is an example of such a molecule. Amino acids contain at least one carboxylic acid group and one amino group. The side chain of the amino acid may also be an acidic or basic group. Your goal in this lab is to identify the amino acid present in the solution you are given.

6.2. Procedure

6.2.1. Titration of water to pH 12

1. Calibrate the pH meter you intend to use.

2. Pour 40 mL of water into a 100 mL beaker.

3. Place the beaker on a stir plate. Rinse a small stir bar and place the stir bar in the beaker.

4. Prepare a buret containing approximately 40 mL of the

titrant, 1 M KOH.[1] Adjust the location of the buret so that the end of the column is an inch or two above the beaker contents.

5. Prepare a pH meter to evaluate the pH of the solution during titration (see Appendix E) – ensure that the end of the pH probe is immersed in the solution on the side opposite the side where additions of titrant are made. Position the pH probe tip well above the stir bar (see Figure 6.2), ensuring the stir bar is unlikely to collide with the probe tip.

6. Follow the titration[2] from its starting point at least past pH 12.5:

   (a) record the pH and buret volume (remember that the bottom of the meniscus (Figure 6.3) is used to measure pH)
   (b) add approximately 1 mL of 1 M KOH (*unless* the change in pH with the previous addition was greater than approximately 0.2 pH units; in this case, add 0.2 mL instead)
   (c) go back to step (a)

6.2.2. Titration of a solution of an unknown amino acid

1. If you have not yet calibrated the pH meter you intend to use, do so.

2. Record which unknown was analyzed: __

3. Rinse the 100 mL beaker you were using and add 40 mL of the unknown amino acid solution to the beaker.

---

[1] KOH is a strong base; avoid handling a concentrated KOH solution carelessly.
[2] Tip: plot the data as the titration proceeds

Figure 6.1. A buret

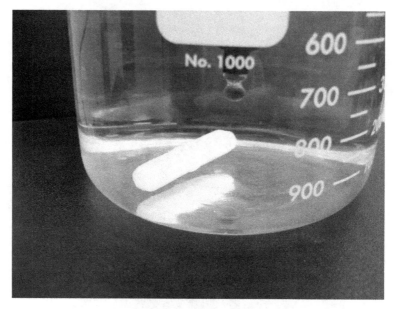

Figure 6.2. pH meter setup

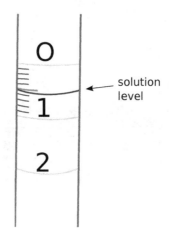

Figure 6.3. The meniscus in a buret

4. Titrate the solution of the unknown amino acid as described above (by repeatedly recording the pH and buret volume and then adding a volume of base).

6.3. Laboratory report

The lab report for this exercise should be prepared following the standard format. Organize your data into a table if you did not do so while recording the data. Graph the data for each titration. Suggest explanations for the differences between the two titration curves. Use your data to generate a hypothesis regarding the identity of the unknown zwitterion. Draw the theoretically possible ionic forms of that amino acid and indicate which form, if any, predominates at pH 1, pH 6, and pH 11. Do any of the forms not exist in aqueous solution, irrespective of pH? Identify the pK_a values of the amino acid in your unknown solution – compare your estimates with the values published in the literature. Consider carefully how you will calculate/estimate the pK_a values (e.g., using first/second derivative values for the curve or using equivalence points).

6.4. Analyzing the data

One of the main objectives you will have while analyzing the data is to determine the pK_a values for the unknown. When graphing pH versus volume of titrant, you may wonder if there is a tool available to aid you in identifying the regions of the graph with lowest slope values. One such tool is SciDAVis (see Appendix F). The "Analysis" features of SciDAVis include a "Quick Fit" function for curve fitting as well as "Differentiate" and "Integrate" functions. Another tool with powerful curve fitting capabilities is R (R provides quite a few models (e.g., those associated with the lm and nls functions)) as well as excellent plotting capabilities (e.g., in this case, the ggplot function and the stat_smooth functions may provide nice visuals of the fits).

6.5. Modeling a titration

In an appendix to the laboratory report, document the use of a spreadsheet program[3] to model the titration of a monoprotic acid with a pK_a equal to the estimated pK_a of the carboxylic acid function of the unknown amino acid you evaluated experimentally (e.g., Figure 6.4). Ensure you include a figure of the spreadsheet in the appendix.

The model should assume that every molecule of the titrant, potassium hydroxide, added acts as a 'strong base', completely dissociating to ^-OH and K^+ in solution.

The titration of HA begins with a 1 L aqueous solution of 0.202 M HA at pH 1.0.

The titration continues with successive additions of 0.02 mole of KOH using a 1M KOH solution. Your spreadsheet model should account for changes in the volume of the solution due to addition(s) of titrant.

In what ways does the plot resemble or not resemble the experimental titration data collected? If the plot differs from the experimental data, suggest possible explanation(s) for the difference or differences.

Can the Henderson-Hasselbalch equation be used to calculate pH for any set of $[A^-]$ and $[HA]$ values? Can the equation be used to calculate the pH of a 1 L solution prepared by adding 0.1 mole of sodium acetate to water?

---

[3]Appendix C describes some of the available spreadsheet software.

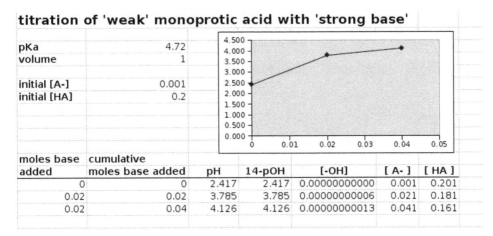

titration of 'weak' monoprotic acid with 'strong base'

pKa	4.72					
volume	1					
initial [A-]	0.001					
initial [HA]	0.2					

moles base added	cumulative moles base added	pH	14-pOH	[-OH]	[A-]	[HA]
0	0	2.417	2.417	0.00000000000	0.001	0.201
0.02	0.02	3.785	3.785	0.00000000006	0.021	0.181
0.02	0.04	4.126	4.126	0.00000000013	0.041	0.161

Figure 6.4. A spreadsheet model of a titration. A screenshot of a spreadsheet model of a titration showing the rows corresponding to the first few additions of base.

Unless your instructor indicates otherwise, submit your "spreadsheet file" with the completed model as an email attachment.

6.6. Resources

Most undergraduate biochemistry texts include an overview of acid/base chemistry. In addition, the Merck Index contains pK_a values for many compounds. "An Introduction to Aqueous Electrolyte Solutions" contains a detailed discussion of acid-base chemistry, including derivations of equations and worked problems. [27]

6.7. Acknowledgments

A similar exercise is described by B. Ganong. [11]

7. Structure visualization I

7.1. Introduction

Two experimental techniques, nuclear magnetic resonance (NMR) and X-ray crystallography, are commonly used to determine the structure of large biomolecules such as proteins as well as the structures of smaller compounds such as the one we are examining in this exercise. The structural description gleaned from such experiments is often stored in a public repository. Generally, the structure data is accessible as a file in the PDB format. Two points of access to PDB structure data NCBI Structure (https://www.ncbi.nlm.nih.gov/structure) and the RCSB Protein Data Bank (http://www.rcsb.org).

The computer is a powerful tool for visualizing molecular structure. This exercise is the first in a series of exercises which will introduce you to molecular visualization software. Appendix G contains descriptions of a number of computer software packages for molecular visualization; many of these programs can use PDB files as input, processing the data in the file to create a pseudo-three-dimensional image of the biomolecule.

In this exercise, you will focus on the structure of glycyl-L-threonine dihydrate, evaluating inter- and intramolecular bonding using the computer as a tool. The course instructor will describe how you can access the PDB files, **gt.pdb** and **test.pdb**, which contain representations of the structure. You may find it convenient to save these files on your computer before beginning the exercise.

7.2. Procedure

7.2.1. opening and viewing the PDB file

1. Review the brief guide to using VMD (see Appendix K).

2. Start the VMD program.

3. Open the gt.pdb structure file.

4. VMD has several different ways to modify the view of a structure. Invest a few minutes in becoming more comfortable with rotating and scaling.

7.2.2. using drawing and coloring methods

Use the "Graphics" menu button on the "VMD Main" window to select "Representations". This brings up a window which allows you to modify the coloring scheme and also the physical manner in which atoms and bonds are depicted (see the "modifying the representation of a molecule" section in Appendix K).

To get a sense of what the drawing methods[1] correspond to, try several of the different drawing methods, including VDW, licorice, dotted, and cartoon. For a familiar "ball and stick" representation, set the drawing method to CPK.

Try several of the coloring methods.

Set the coloring method to Name. In this coloring scheme, each atom receives a different color based on its identity in the periodic table. One can determine which atoms correspond to which color by using the Color Form (go to the "VMD main" window, select the Graphics pull-down menu, and select Colors).

---

[1]You may wish to reference the VMD documentation to determine what a method corresponds to in terms of physical representation

atom	color
H	white
O	red
N	blue
C	cyan[2]
S	yellow
P	tan

Table 7.1. Default VMD color assignments

7.2.3. characterizing bonding in the structure

Do any elements appear to be missing from the structure? If you look closely, you should see that there are several oxygen atoms "floating" without anything apparently bonded to them. This is because, in many cases, the resolution of NMR and X-ray crystallography structures is insufficient to identify the locations of hydrogen atoms. Software packages are available that will make "educated guesses" regarding hydrogen locations.

To visualize such a modified structural data set (in this case, the locations of the H's were estimated using a program called openbabel), load the test.pdb file.

To begin to characterize the bonding patterns in the crystal structure, it may help to first examine a single glycl-L-threonine molecule and the molecules in its immediate vicinity. Use the "Selections" tab in the "Graphical Representations" window to select a single 'chain' (use chain A) to view. Note that, in VMD, a 'chain' is an arbitrary label for a set of atoms (e.g., in proteins with quaternary structure, each polypeptide is typically distinguished as a distinct VMD chain). You should see a representation of a single glycyl-L-threonine molecule. **Print this view out.**

> note: If the image background is black, invert colors and print with a white background in order to make annotating the image easier.

On the same sheet of paper on which you printed the view in the above step, draw, by hand, the structure of glycl-L-threonine using the conventional method for representing organic compounds. On the VMD CPK representation of glycyl-L-threonine, label the

glycine side chain
threonine side chain
carboxy terminus of the dipeptide
amino terminus of the dipeptide

Color chain A a unique color by selecting the "Color ID" or "Beta" coloring method (under the 'Draw style' tab in the 'Graphical Representations' window) and change the drawing method for chain A to "Surf".

Create a second representation of test.pdb (use the "Create rep" button in the "Graphical representations" window and, for this representation, set 'selected atoms' to 'all', 'coloring method' to 'Name', and the 'drawing method' to 'CPK'). This view should show a "space filling" model of chain A surrounded by ball/stick views of the molecules adjacent to chain A. Try rotating this view to get a better sense of the spatial distribution of the other molecules relative to chain A. In the next section, we will examine the nature of these interactions.

7.2.4. identifying residues and measuring distances

Identify a water molecule in the vicinity of chain A.[3] Visually identify a water molecule which you anticipate is interacting with chain A. Now, ensure that you are able to identify residues/regions of the different molecules displayed (see

---

[3]If the plethora of atoms surrounding chain A seem a bit overwhelming, one can direct VMD to display only water molecules surrounding chain A: edit the representation which currently has selected atoms set to "all" so that the selected atoms value is "resname HOH".

the "identifying residues or regions of a molecule" section in the "brief guide to using VMD"). **Unambiguously identify the water molecule you identified by recording the index numbers of the H and O atoms.**

Make a table with the following column labels:

> index number of atom in chain A
> name of atom in chain A
> index number(s) of atom(s) interacting with atom in chain A
> type of interaction
> distance between the two atoms
> typical distance for this type of interaction

Focus on the water molecule you chose first. **Use the first row of the table you just created to describe one interaction that appears to be occurring between the water molecule and chain A.**

- identify the atom(s) in chain A which appears to be interacting with the water molecule (what is its index number? what is its name?)

  > note: At this point, it may be helpful to switch the chain A coloring method to "Name".

- identify the corresponding atom(s) in the water molecule which appears to be interacting with the atom(s) just identified in chain A

- determine the distance between the two atoms

- identify the type of bonding interaction you think is occurring

- do some research:

  - what is the expected distance between the bonding atoms in the type of bond interaction you think is occurring?

  - ensure you cite the source of the bond length data

- does the reference value match the observed distance?

Identify another glycyl-L-threonine molecule which appears to be interacting with chain A. Use the procedure you just used with the water molecule to evaluate the interaction between chain A and the other glycyl-L-threonine molecule you have chosen (i.e., fill out a second row in the table).

7.2.5. extra credit

When the 'test.pdb' file was generated by openbabel, it didn't do a perfect job of extrapolating the positions of the missing hydrogens. Take a look at the structures of a glycyl residue and a threonyl residue and see if you can identify a structural error that openbabel did not correct.

7.2.6. optional additional fun: using software to evaluate intermolecular contacts

If you are curious about your guesses, try using MolProbity[4] to analyze the pdb file. molprobity is essentially an online interface to the "reduce" and "probe" programs (if you have a Linux platform, you can use these tools directly by installing them on your computer). With these tools, you can produce a "kinemage" which shows clashes between groups, H bonding, and van der Waals interactions. SPDBV pock has tools for seeing H bonds as well.

7.3. Laboratory report

Do not prepare a formal laboratory report (i.e., with "Introduction", "Methods", "Results", and "Discussion" sections).

---

[4]http://molprobity.biochem.duke.edu

Instead, prepare a report which addresses all items in bold in the above exercise.

7.4. Acknowledgments

Inspiration for this exercise was an exercise designed by A. Glasfeld. [13]

8. Structure visualization II

8.1. Introduction

In this lab, online protein sequence retrieval and analysis utilities are used to characterize members of the myosin superfamily of proteins. The lab report will not follow the standard report format; rather, it should contain the data obtained over the course of the exercise along with answers to the questions posed below. Ensure you include an answer to all questions or directives in the exercise below in your report.

8.2. Procedure

8.2.1. retrieving a protein sequence

The *Homo sapiens* MYH15 gene is located on chromosome 3. This gene encodes the myosin-15 protein. Retrieve the *Homo sapiens* myosin-15 protein sequence (avoid sequences that are described as "predicted" or "hypothetical"). Record the accession number of the sequence you obtained.

NCBI (https://www.ncbi.nlm.nih.gov/) provides interfaces for retrieving many types of data. You may wish to begin your search for the polypeptide sequence either through the "Gene" interface or through the "Protein" interface. Note that searches can be limited using "search field tags". For example, the [ORGN] tag is used to limit a search to a particular organism name, the [TITL] tag searches the title record (a standard title field will include the organism, product name, gene symbol, molecule type and whether it is a partial or complete coding sequence) for indicated keywords, the [ALL] tag searches all fields for the indicated keyword, and the [ACCN] searches the accession number (a unique code assigned to each entry in the database(s)) field. The interface also accepts logical arguments using AND, OR, and NOT. Together, these are a powerful means of limiting and focusing searches. For example, a search for "eye[ALL] AND homo sapiens[ORGN]" would look for all entries for humans with the word "eye" in the record.

8.2.2. an analysis of the primary sequence

How many amino acids in length is the protein sequence retrieved?

Based on this, what is the predicted molecular mass of the protein?[1]

Compare the size of this protein to that of two other well-known *Homo sapiens* proteins: the mature (after post-translational processing) α-actin monomer and the glycolytic enzyme aldolase A. Does myosin-15 seem to be a relatively large or relatively small protein?

8.2.3. secondary structure prediction

Regions of secondary structure can be predicted using computer algorithms. Use a computer algorithm such as nnpredict (see Appendix H) to predict regions of secondary structure in myosin-15. Briefly describe the results of this analysis in prose, clearly indicating the locations of regions of predicted secondary structure in the myosin-15 protein. Note

---

[1]Ensure your answer is explicit with respect to units of mass.

that you are not required to use nnpredict.[2] Ensure you reference or cite the tool used.

> you may find it easier to change the "Display" option in the search result window to ASN.1 or another format which displays the primary sequence without intervening numbers.

8.2.4. fold recognition

A protein is typically organized into one or more **domains**. A domain, a region of the polypeptide chain which independently folds into a stable tertiary structure, is the fundamental unit of tertiary structure. Often, a protein domain has a distinct known function such as an ATPase activity or a ligand binding activity. Given the large amount of sequence information available, domains can be tentatively identified based on primary sequence data using computer analysis. For example, chymotrypsin, urokinase, factor IX, and plasminogen all have a distinct serine proteinase domain (a 245 amino acid segment arranged into two domains). Use a computer tool such as SMART (see Appendix H) to predict whether myosin-15 contains any known domains. Ensure you reference or cite the tool used. List the name and location (amino acid residue numbers) of each predicted domain. Provide, in prose, a brief description/summary of the biological function of each domain.

---

[2]Other secondary structure prediction algorithms are available; some have interesting names like 'GOR IV' and 'PREDATOR'.

8.2.5. experimental observations regarding myosin 15

Earlier you calculated the predicted molecular mass of myosin 15. Perform a literature search to determine whether the actual molecular mass of myosin 15 has been experimentally determined. In addition, provide a brief summary of what is known regarding the cellular and physiological roles of myosin 15. For example, is a specific cellular function of myosin 15 established? Is a particular disease or syndrome linked to myosin 15?

You have used computer programs to predict secondary and tertiary structure elements of myosin 15 based on the primary sequence of the polypeptide. Has the tertiary structure of myosin 15 been experimentally determined? If so,

1. list the reference

2. download the PDB file and, using VMD or a similar program, compare the predicted arrangement of secondary structure elements wit the observed (experimental) tertiary structure

> You may want to save a copy of your completed lab report for your own use during the following week's laboratory exercise.

8.3. Laboratory report

The lab report for this exercise should be submitted as a written report, with printed copies of the data from the exercise. The report should not be written with an "Introduction", "Methods", "Results", or "Discussion" section. Rather, the report should chronicle all the results of the exercise and answer all explicit questions in the exercise.

8.4. Resources

Take advantage of online tools for this assignment. For example, the protparam tool is useful for primary structure analysis. For links to this and other, related resources, see Appendix H, Appendix I, and Appendix J.

9. Tertiary structure prediction

9.1. Introduction

In this exercise, the goal is to design a simple polypeptide which consists of two alpha helices connected by a short loop such that the polypeptide adopts a helix-loop-helix structure.

Some portions of this exercise are computing-intensive and may require hours to days of computing time to complete (i.e., it is probably best to start this lab at least 3-4 days in advance of the due date).

Please do not abuse the public servers for tertiary structure prediction. *Especially for ab initio predictions*, these 'jobs' are computationally expensive. For these servers,

- only submit one job at a time

- do not submit more than one job per day

9.2. Procedure

1. Design (on paper) an oligopeptide of 40 to 60 amino acid residues in length which you anticipate will assume an α-α (helix-loop-helix) structure such that the two α-helices interact with each other closely along the length of the

helices with the interaction stabilized primarily by "hy-drophobic interactions". Briefly describe the rationale behind your design and record the primary sequence of your initial polypeptide.

> Consider using a helical wheel tool such as Pep-Wheel (EMBOSS) or drawing a wheel.

2. A number of different algorithms have been developed to make secondary structure predictions based on the primary sequence of a protein. Use nnpredict or a similar secondary structure prediction tool[1] to estimate the predicted secondary structure of your oligopeptide. Is the predicted secondary structure similar to your expected secondary structure? If not, consider revising your sequence before using a tertiary structure prediction tool.

3. Choose a tertiary structure prediction tool capable of predicting a *complete* tertiary structure (i.e., capable of outputting a PDB file describing the generated model) based on primary sequence. Note that your instructor may have specific recommendations regarding which servers you should consider.

   - Record the name, and a complete citation, for the tertiary structure prediction tool you used

   - Briefly describe the methodology this software employs to predict tertiary structure

   - Save the predicted structure as a PDB file (some servers may call this a "Rasmol" file)[2]

---

[1] for more on nnpredict and other computational tools for protein structure, see Appendix H

[2] If, for some reason, you are unable to obtain a tertiary structure prediction, you must document that you attempted to obtain a prediction with at least two services. In addition, you must describe, in detail, the algorithms used by those services and the probable reason you did not obtain a prediction. Finally, you must locate a protein with a helix-loop-helix domain and conduct the analysis described below

- E-mail the instructor a copy of the PDB file

- Use a protein structure viewing utility (VMD is recommended) to examine the PDB model of the structure

- Print a picture of the structure using a "ribbon" or "cartoon" drawing method. Label the N-terminus and the C-terminus. Label any regions of secondary structure.

- Is the predicted structure similar to your target structure? Identify four solvent-exposed side-chains (identify each by circling it in the figure and labeling it with the residue name and residue number). Are these side-chains which you anticipated would be solvent-exposed when you initially designed the polypeptide?

- If the model contains a helix-loop-helix structure, inspect the model to determine whether the interactions stabilizing the helix-helix interaction are indeed hydrophobic interactions. Use VMD to show two residues (use the CPK method for visualization) interacting in this fashion; identify and label (in the figure) each residue in terms of its amino acid side chain and its numeric position in the polypeptide.

- Identify by name and number the amino acid residues, if any, which appear to be interfering with the desired folding of the structure. Suggest modifications, and the rationale behind them, which you believe will improve things.

4. It is often of interest to evaluate the degree of similarity between two protein or nucleotide sequences. One program used to evaluate similarity between two known nucleotide or protein sequences is CLUSTAL. However, often it is of interest to compare a known polypeptide sequence against a database of hundreds or thousands of

---

using that domain.

other known polypeptide sequences. A powerful tool to compare a protein or nucleotide sequence against those in the sequence databases is BLAST. Use BLAST[3] to confirm that the novel sequence you have designed is indeed novel (the criterion we will use in this exercise is an E value greater than 0.01). Print out at least the first few lines of your BLAST results and submit them with the laboratory report.

9.3. Laboratory report

The lab report for this exercise should be submitted as a written report, with printed copies of the data from the exercise. The report should not be written with an "Introduction", "Methods", "Results", or "Discussion" section. Rather, the report should chronicle progress through the exercise, including results for each section of the exercise as well as answers for all explicit questions in the exercise.

9.4. Additional resources

See Appendix J.

---

[3]How does one interpret BLAST results? One general way to approach such questions is to read the online "Help" and "FAQs" sections, etc. However, the discerning user may notice that, if he or she is using www.ncbi.nlm.nih.gov/BLAST, the user's BLAST results appear in a window in several different formats. Beneath a graphical representation of the results is a list of protein names. To the right of the list of names are two columns, "SCORE (the S value)" and "E VALUE". These two columns indicate the degree of similarity between the query sequence and the sequence whose name is in the first column. A low E value indicates a high degree of similarity between the two sequences. To actually see each alignment, scroll down further in the results page. You will find each alignment depicted along with a summary of information describing the alignment.

10. Reading a scientific paper

Reading a scientific paper is generally a challenging experience quite different from other modes of reading. If you are unsure which scientific paper is to be used for this assignment, obtain that information from the course instructor before proceeding.

On a single sheet of paper (single-spaced, 12 pt or greater font), review the paper, addressing the following questions.

1. What is the background behind the study?

   - do there exist related studies?

   - is there a compelling motivation for the study (e.g., human health concerns, etc)?

2. What overarching question(s) does the paper answer? In other words, what hypothesis is addressed in the paper?

3. What conclusion(s) is/are drawn with respect to the hypothesis(es) addressed by the paper?

4. Summarize the **evidence** supporting the conclusion(s) drawn by the author(s), the **quality of that evidence**, the **quality of the logic** used in drawing conclusions from the data, and **your conclusion** (i.e., whether you believe the evidence does or does not support the hypotheses).

On a separate sheet of paper, answer the following questions:

1. What methodology is employed to test the hypothesis or hypotheses? (answer this question in detail on a separate page)

2. Address the quality of the writing: is the paper poorly written?

   - is the data poorly presented or presented in an unclear fashion?

   - do the authors omit important content?

   - are there grammatical or spelling errors?

?? contains a helpful overview describing the process of reading a scientific paper and some of the stumbling blocks that may be encountered.

11. Enzyme kinetics

11.1. Prelab

This laboratory session **requires** advance planning. You may find it helpful to draw out a flowchart in advance describing the procedures you will use.

Complete the following steps, along with the exercise and questions in the 'Introduction' section, prior to the lab session:

1. Write the Michaelis-Menten equation. Define each variable or constant in the equation.

2. List the experimental conditions which are assumed when deriving the Michaelis-Menten equation.

3. Given a sample with high α-amylase activity, what color would you predict would be observed upon evaluation of the sample using the α-amylase assay described below?

4. Given a Michaelis-Menten enzyme with K_m of 1 µg mL^{-1} and a V_{max} of 30 µg mL^{-1} min^{-1}, predict the velocity of the reaction if the enzyme is incubated with 30 µg mL^{-1} substrate.

11.2. Introduction

In this experiment, your interest is in characterizing the activity of an enzyme, α-amylase, in a crude sample (saliva) by estimating the V_{max} and K_m of the enzyme.

For many enzymes, the reaction rate is expressed as conversion of one µmole substrate per minute or per second. The related measure of catalytic activity is typically expressed as an "International Unit" (IU), defined as 1 µmol substrate/(min * L). The IU, although commonly used, is not a SI unit. Use of the IU unit is discouraged in favor of the corresponding SI unit, the katal.[1] One katal is defined as the amount of enzyme which converts one mole of substrate per second.[2]

However, the system we are considering isn't quite typical in at least two respects:

1. the substrate is a polymer; thus, measuring consumption of moles of substrate per unit time does *not* give us an indication of the velocity of the reaction in less we clearly define the substrate in a suitable manner; however, we can measure the consumption of starch in µg starch per unit time

2. since the enzyme is not purified and the absolute concentration of α-amylase in the saliva is unknown, a V_{max} expressed relative to µg or µmole of enzyme cannot be calculated. However, a V_{max} expressed in terms of the 'concentration' of saliva can be calculated. Furthermore, it should be possible to estimate the apparent K_m of salivary amylase.

A protocol for a colorimetric α-amylase assay is given be-

---

[1] However, if the 'katal' is mentioned to the average biochemist, there is a significant probability that the response will be a blank stare.

[2] The transition to the katal is described in "The Tortuous Road to the Adoption of katal for the Expression of Catalytic Activity by the General Conference on Weights and Measures" (René Dybkær. Clinical Chemistry 48: 586-590, 2002).

low. The method is based on that described by Somogyi. [24] The 10-fold concentrate of the iodine solution used in the experiment is a red-brown solution which contains 2% iodine and 0.002N KI in the same buffer as is used for the starch solution. The triiodide anion present in the iodine solution will interact with starch (if starch is present) to form a dark blue complex with an absorbance maximum at approximately 620 nm.

It is important to understand, *in advance*, the conceptual steps of this laboratory. In order to estimate K_m and V_{max}, one must evaluate v_o across a range of substrate concentrations for some fixed enzyme concentration. Both the range of substrate concentrations and the enzyme concentration to be used should be optimized.

Ideally, the range of substrate concentration includes both the pseudo-linear and asymptotic regions of the Michaelis-Menten data (v_o vs. [S]). Consider Figure 11.1. If one only evaluated a range of [S] greater than 5, one might obtain a good estimate of V_{max} but the estimate of K_m would be relatively poor.

The second concern, the enzyme concentration used, will be addressed in the first part of this lab. Does the enzyme concentration 'make a difference'? Under Michaelis-Menten conditions, what is the answer?

Using a graphing program, plot v_o vs. [S] for substrate concentrations equal to 1, 2, 4, 8, 16, and 32 μM where $V_{max}=0.001$ μmol/min. How does the plot differ from the corresponding plot with V_{max} equal to 1? to 10000? What do you conclude about the effect of changing the concentration of enzyme under Michaelis-Menten conditions? Are there any practical problems you envision with respect to collecting data for any of the three plots just prepared?

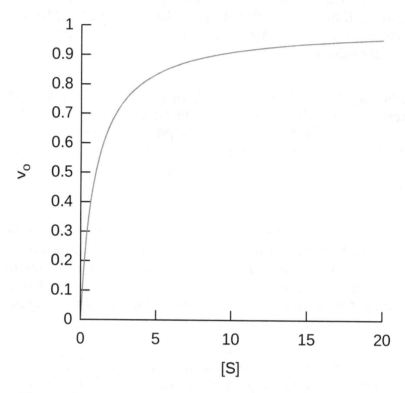

Figure 11.1. Relationship between v_o and [S] for a Michaelis-Menten enzyme. The relationship of v_o and substrate concentration is shown for a hypothetical Michaelis-Menten enzyme with a K_m of 1 and a V_{max} of 1 where the units are arbitrary.

Sample	µL starch solution	µL Tris-phosphate buffer	µL 0.5% NaCl	µg starch/µL	A_{620}
A	139.5	0.5	10		
B	69.8	70.2	10		
C	34.9	105.1	10		
D	17.4				
E	8.7				
F	4.3				
G	0				

Table 11.1. Standard curve data. Some lines are partially filled in as examples.

11.3. Procedure

11.3.1. Construct the standard curve

The standard curve describing the relationship between the concentration of starch in the sample[3] and A_{620} will be used in subsequent steps.

Use the table below as a starting point for collecting the data for the standard curve.

Prepare the following 150 µL samples using the 0.75 g/L starch solution:

1. 139.5 µL starch solution, 10.5 µL NaCl solution

2. 69.8 µL starch solution, 69.8 µL dilution buffer, 10.5 µL NaCl solution

3. 34.9 µL starch solution, 104.6 µL dilution buffer, 10.5 µL NaCl solution

4. 17.4 µL starch solution, 122.1 µL dilution buffer, 10.5 µL NaCl solution

---

[3]by "the sample", we mean the sample from which 150 µL are removed and mixed with 600 µL of the iodine solution in the assay

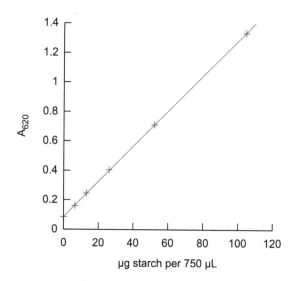

Figure 11.2. A standard curve for the α-amylase activity assay

5. 8.7 µL starch solution, 130.8 µL dilution buffer, 10.5 µL
   NaCl solution

6. 0 µL starch solution, 139.5 µL dilution buffer, 10.5 µL
   NaCl solution

Add 600 µL of the 1X iodine solution (0.0002 N KI, 0.2%
I_2) to each 150 µL sample and mix the contents of the tube
directly before reading the A_{620}.[4]

Record the data collected above as a table and construct a
graph of the data as a standard curve (see Figure 11.2 for
an example of a standard curve). Include a table containing
the numeric data along with the corresponding graph in the
laboratory report for this project.

---

[4]If you are not familiar with the basics of using a spectrophotometer,
consider investing some time in reviewing Appendix P.

11.3.2. Preparing saliva and starch solutions

As you conduct preliminary experiments to determine which enzyme and substrate concentrations are appropriate, prepare enzyme dilutions in the following manner:

1. Place at least 1 mL of saliva in a 1.5 mL centrifuge tube and centrifuge at high speed for at least 30 s. If there is a large amount (greater than 20% of the total volume) of yellow-brown 'crud' at the bottom of the tube, discard that sample and obtain more saliva from the same individual.

2. Transfer the supernatant to a new tube labeled "100% SAL".

3. Label a new microcentrifuge tube indicating the intended dilution and perform the dilution using 0.5% NaCl. For example, for a 20% saliva solution, label a tube "20% SAL" and then add 800 µL of 0.5% NaCl and 200 µL of the saliva supernatant to the 20% tube. Vortex vigorously for at least 10 s to ensure the tube contents are well-mixed.

The stock starch solution is 0.75 mg/mL. Prepare more dilute starch solutions by diluting in Tris-phosphate buffer. For example, to prepare a 0.375 mg/mL starch silution, add 5 mL Tris-phosphate buffer to 5 mL of the stock starch solution; mix thoroughly.

11.4. α-amylase activity assay

For each sample, have at hand:

- two microcentrifuge tubes
- a cuvet suitable for a 1 mL sample volume

In addition, have at hand:

- a 200 µL variable volume pipettor

- 200 µL pipet tips

- a 1000 µL variable volume pipettor

- 1000 µL pipet tips

- a vortexer

- a spectrophotometer

1. Aliquot 600 µL of the 1X iodine solution[5] into a 1.5 mL microcentrifuge tube (one can also use a 100x13 or 12x75 glass tube)

2. Prepare the reaction mixture:

   - withdraw 1.2 mL of the substrate solution[6] and transfer the solution to a 1.5 mL microcentrifuge tube

3. Begin the assay:

   - add 85 µL of the sample of interest to the 1.5 mL tube containing the starch solution

   - immediately mix well (vortex for 2 to 3 s or invert several times) and note the time

   - at 3 min, remove 150 µL from the starch/sample tube and add the 150 µL aliquot to a tube with iodine solution; mix (vortex or invert several times) the iodine tube/sample

   > The assay can be performed at 30°C or 37°C instead of room temperature but the time should be modified appropriately.

   - evaluate the absorbance:

---

[5]"10X" stock iodine solution: 2% iodine, 0.002N KI in reaction buffer (store in the dark, tightly capped, at room temperature)

[6]The stock starch solution consists of 0.75 g/L soluble starch in 20 mM Tris-phosphate, 10 mM NaCl, 1 mM NaF, 0.1 mM NaN_3, pH 7.0. It should be shaken well before use.

blank on water

measure A_{620} for the sample

11.5. Laboratory report

This exercise is only one component of a multi-week set of exercises in which you will characterize various characteristics of the enzyme α-amylase. The lab report for this exercise should be prepared following the standard format. Since this is just one component of a larger project (purification and characterization of human salivary α-amylase), the report should be prepared after the final α-amylase laboratory session and should be a comprehensive report describing the materials, methods, and collected data from all the α-amylase lab sessions.

From this week's data, one can begin to evaluate some characteristics of α-amylase. What is the apparent K_m of the enzyme? What is the apparent V_{max} of the enzyme? Consider beginning doing the research you will need to do to have a well-written introduction and discussion. For example, have others already characterized the kinetics of human salivary α-amylase? Are K_m and V_{max} estimates already in the scientific literature? How do your estimates compare to these published values?

12. Protein purification

12.1. Prelab

1. Draw a flowchart describing the purification process. Figure 12.1 shows a fragment of such a flowchart.

2. Using chemical structures, draw the reaction catalyzed by α-amylase.

12.2. Introduction: purification of α-amylase

In this and subsequent laboratory sessions, you will purify and characterize α-amylase (1,4-α-D-glucan-glucanhydrolase), a starch hydrolytic enzyme which catalyzes the hydrolysis of α-linked glucose polymers. α-amylase is found in secretions such as the saliva and in tissues such as the liver of mammals. α-amylase is also found in plants and in bacteria.

Session 1: α-amylase purification
Session 2: evaluation of protein concentration of each fraction
Session 3: α-amylase activity assay
Session 4: column chromatography of α-amylase
Session 5: SDS-PAGE of purified α-amylase

12.3. Session 1: α-amylase purification

This purification protocol is based on that described by Schramm and Loyter [23].

Reagents provided:

- 95% ethanol

- 0.2M phosphate buffer, pH 8.0

- glycogen solution (2% in water)

1. In advance, prepare a flowchart describing the purification process (an example of how such a chart might be started is below).

2. Collect at least 3 mL saliva (a 15 mL disposable centrifuge tube works well for this). Mix well and distribute into two 1.5 mL microcentrifuge tubes.

3. Remove debris and particles from the sample:

   - balance the two 1.5 mL tubes in the microcentrifuge

   - centrifuge at 8000 g[1] for 5 min.

4. From each of the 1.5 mL microcentrifuge tubes, remove 960 µL of supernatant and transfer the 960 µL to a 2 mL microcentrifuge tube (you will need two 2 mL microcentrifuge tubes – one corresponding to each of the two 1.5 mL microcentrifuge tubes).

   note: The pellet is disturbed relatively easily.

5. Combine the two 960 µL volumes into a single volume in one of the 2 mL tubes; mix thoroughly. Divide the 1920 µL back into two 960 µL volumes (960 µL in each of the two 2 mL tubes).

---

[1]9300 rpm if using a Fisher Scientific Marathon Micro A microcentrifuge

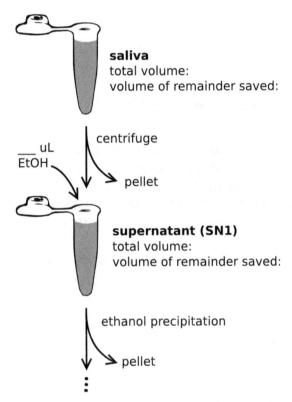

saliva
total volume:
volume of remainder saved:

___ uL
EtOH

centrifge

pellet

supernatant (SN1)
total volume:
volume of remainder saved:

ethanol precipitation

pellet

Figure 12.1. Sample segment of purification flowchart

6. Confirm that you got it right: at this point you should have two 2 mL microcentrifuge tubes, each containing 960 µL of the supernatant.

7. Pool and save the remainder of the supernatant (you can label this sample 'SN1') in a 1.5 mL tube; keep this sample on ice.

8. After this step, keep the reagents used in all steps below 4°C (i.e., keep the reagents on ice).

12.3.1. Ethanol precipitation

1. Adjust each 960 µL sample to 40% ethanol: add ice-cold ethanol drop-by-drop, mixing the tube contents after addition of each drop (this can be done by placing the tube on a vortexer and adding the ethanol while vortexing). After addition of the last drop, mix thoroughly. Centrifuge the tubes at 10000 g[2] for 10 min.

> If adding 100% ethanol, 0.4 = x µL / (960 µL + x µL)

2. Obtain two new (empty) microcentrifuge tubes. These tubes should be clear-bottomed with no seams to facilitate easier visualization of the pellet. Label each tube "EP". Place two tubes on ice.

3. Transfer 1 mL of supernatant from each tube centrifuged in the previous step to each of the 1.5 mL microcentrifuge tubes labeled "EP".

4. Save the remainder of the sample (you can label this sample 'SN2') on ice.

---

[2]10500 rpm if using a Fisher Scientific Marathon Micro A microcentrifuge

72

12.3.2. Ethanol/glycogen precipitation

1. After the supernatants have cooled for 1 min, add to each tube labelled "EP" (in the order listed)

   - 60 µL 0.2M phosphate buffer, pH 8.0

   - 50 µL glycogen reagent OR 40 µL glycogen reagent[3] and 13 µL "GlycoBlue" ("GlycoBlue" is a dye-labelled glycogen at 15 mg/mL))

   - 80 µL 95% ethanol

2. Mix the contents of each tube thoroughly. Incubate on ice 5 min. Centrifuge at 5000 g[4] for 3 min.

3. Label a new 15 mL centrifuge tube 'SN3'. Carefully decant the supernatant from each tube into the 15 mL centrifuge tube and save the pooled supernatant.

   > note: The pellet may be very small and clear (i.e., difficult to see) if GlycoBlue isn't used.

12.3.3. Wash the precipitate

1. Ensure that, in advance, *you* prepare the resuspension solution (750 µL H_2O, 72 µL 0.2M phosphate solution, pH 8, 600 µL 95% ethanol)

2. Resuspend the pellets in 1 mL (total final volume) of ice-cold resuspension solution in the following manner: add the 1 mL to the first tube, flick or vortex to resuspend the pellet; transfer the entire solution into the next tube containing a pellet; flick or vortex vigorously to resuspend the pellet.

---

[3]glycogen reagent: 2% oyster glycogen (Sigma Type II) in water (it is important to centrifuge this at 11000 g for 5 min directly before use)
[4]7400 rpm if using a Fisher Scientific Marathon Micro A microcentrifuge

> - Remove 100 µL from this tube and transfer to a new tube labeled "RES"; save this tube on ice

3. Place the tube containing the remainder of the resuspended precipitate in the centrifuge oriented so that the lid hinge is furthest from the centrifuge rotor (this allows you to anticipate the location of the pellet). After balancing the rotor, centrifuge at 5000 g for 3 min.

4. Carefully decant the supernatant into a new 1.5 mL microcentrifuge tube labeled "SN4" and save this tube.

12.3.4. Prepare a final solution of purified α-amylase

Resuspend (vortex or flick the tube) the precipitate in 200 µL 0.01 M phosphate buffer,[5] pH 8.0 (you can label the resuspended precipitate sample 'PUR').

12.3.5. Storing samples for future analysis

For several of the above purification steps, a sample was obtained. If not already in two equivalent tubes, divide each sample (the SN1 sample, the SN2 sample, the RES, sample, and the PUR sample) into two tubes in equal volumes. For each sample, store one tube at -20°C and the other at 4°C (α-amylase can be stored for 2-3 weeks at 4°C or one can assay activity immediately).

12.4. Cleanup and sanity checks

1. Use a *biohazard* container to dispose of the tubes associated with steps prior to the ethanol precipitation steps.

---

[5]You need to prepare this.

2. Have you carefully labeled one tube from each pair and stored it at 4°C?

3. Have you carefully labeled one tube from each pair and stored it at 20°C?

12.5. Laboratory report

While it isn't a bad idea to begin working on a draft of your laboratory report, the final version of the laboratory report should be prepared after the final α-amylase laboratory session. This report should be a comprehensive report describing the materials, methods, and collected data from all the α-amylase lab sessions.

At this point, you do not have enough data to prepare a purification table describing total activity, total protein, and specific activity for each step in the purification process. In the upcoming labs, you will evaluate total protein and enzyme activity for your samples. From this data, you will be able to construct a table describing each purification step in terms of total activity, total protein, specific activity, and, from these, fold-purification.

12.6. Acknowledgements

A similar exercise was described on the Smith College web site. [3]

13. Assaying protein concentration

13.1. Introduction

In the previous lab session, you (hopefully) purified α-amylase. During the next lab session, you will evaluate the total amylase activity present in the samples representing different stages of the purification process. To calculate specific activity (the amount of activity present per amount of protein present) one must also quantify the amount of total protein present in each of the samples for which activity was evaluated. In this lab session, you will employ a protein concentration assay[1] to estimate the total protein mass and concentration in each of the samples for which α-amylase activity will be evaluated.

13.1.1. In advance

Review the description of the Bradford assay in Appendix N and prepare a table describing the sample volumes for the standards you will use and for the amylase samples.

Begin to prepare a second table, the purification table for the α-amylase purification.

Which of the samples must be evaluated to determine protein content for each purification step?

---

[1]For more on protein assays, see Appendix M.

13.2. Session 2: Determine the protein mass of each fraction

Reagents provided: Bradford reagent, 3 mg/mL albumin

1. Determine which samples you will evaluate. If necessary, return to the previous section.

2. Determine the protein mass of each fraction using the Bradford assay protocol (see Appendix N).

   - data points for standards should be in duplicate

   - sample data points (0.1, 2, and 8) should not be in duplicate

3. Ensure that, as soon as you have removed the appropriate volumes from the tubes used in the purification protocol, that the tubes are returned to storage at the appropriate temperature (4°C or -20°C).

4. α-amylase is relatively stable for months in buffered sodium chloride, pH 7.0, provided the protein concentration exceeds 0.1%. Is the protein concentration in each tube sufficient to ensure that the protein will be stable at 4°C over the course of the coming week?

13.3. Laboratory report

While it isn't a bad idea to begin working on a draft of your laboratory report, the final version of the laboratory report should be prepared after the final α-amylase laboratory session. This report should be a comprehensive report describing the materials, methods, and collected data from all the α-amylase lab sessions. The laboratory report should include the raw data from this session (the A_{450}/A_{590} values as an appendix or attachment or supplementary data section) and a plot of the standard curve and estimates of the protein concentrations in the sample(s) evaluated.

You may also want to start thinking about the "big picture" of the purification. For example, in any purification, it is of interest to determine how much raw material you started with and how much pure product you obtained at the end. Can you use the data you already have to estimate the total mass of α-amylase your purification yielded? to estimate the total mass of protein you started with for the purification?

The specifics associated with the data produced by this assay should also be considered. For example,

- Does the standard curve look "good" or are some data points at locations other than the expected locations?

- What about the variability between the estimated concentration using different sample dilutions? What might contribute to the variability?

Ensure that, in estimating the concentration of your protein sample, you (1) generate a standard curve using a linear or non-linear regression as you deem appropriate, (2) use the curve to determine the protein concentration in each dilution of the sample, and (3) use the dilution factor to calculate the estimated concentration of the original sample.

14. Assaying protein activity

14.1. Prelab

Answer the following questions prior to the lab session

1. Given a sample with no amylase activity, what color would you predict would be observed upon evaluation of the sample using the α-amylase assay described below?

2. Given a sample with high α-amylase activity, what color would you predict would be observed upon evaluation of the sample using the α-amylase assay described below?

14.2. Introduction

This exercise is one component of a multi-lab experiment. In the first lab session, human salivary α-amylase was partially purified. In the second session, the protein concentration was evaluated for each fractions associated with the protein purification. In this session, the α-amylase activity is evaluated for each protein purification fraction. In two upcoming sessions, the efficacy of the purification scheme will be evaluated using column chromatography and SDS-PAGE.

In the activity assay used in this lab session, an iodine-containing solution is used to detect starch. The iodine-containing solution used in the experiment is a red-brown solution

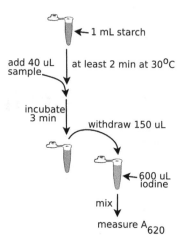

Figure 14.1. The α-amylase activity assay. A flowchart representation of the α-amylase activity assay

prepared as 2% iodine, 0.002N KI. The triiodide anion which forms in this solution interacts with starch to form a dark blue complex.

Along with the iodine solution, one should have on hand the following before beginning the experiment: a 0.75 g/L starch solution, a spectrophotometer, and cuvets.

14.3. Session 3: α-amylase activity assay

For each of the fractions preserved from the purification procedure (Session 1),[1] evaluate the activity of the enzyme:

1. For each protein purification sample, label two microcentrifuge tubes. For each pair of tubes, fill one tube with 600 μL iodine solution.

---

[1]These are the fractions stored at 4°C.

2. In addition, prepare tubes for a "0%" sample (corresponding to the absence of α-amylase: use 40 μL of water as
the sample) and a "100%" sample (corresponding to all
starch consumed: use 40 μL of undiluted saliva as the
sample and incubate for 10 min before evaluating absorbance).

3. Evaluate the activity of each sample (see activity assay
below).

4. Save the original samples in case they are needed in the
future. After this lab, store all samples at -20°C.

14.3.1. assay: for each sample...

> You need to, in advance, plan how you will record the
> time values so that Δt can be calculated.

A. Prepare the reaction mixture:

- withdraw 1 mL of the 0.75 g/mL starch solution and
transfer the 1 mL volume to a 1.5 mL microcentrifuge
tube

- heat the tube (the one you just prepared containing
the starch substrate...) to 30°C for at least 2 min. on
the heat block

B. Begin the assay:

- add 40 μL of the sample[2] to the 1.5 mL tube containing the starch solution

- immediately mix well, transfer to a 30°C water bath
or heat block. This represents the "start time" for

---

[2]Since you don't know the activity in each protein purification, consider,
for SN1 and SN2, starting with a 200X dilution (in 0.5% NaCl in H_2O)
and, for PUR, starting with a 10X dilution to get a sense of range. *Do
not dilute the "0%" or "100%" samples.*

the calculation of v_0.

- at 3 min, remove 150 µL (immediately replace the starch/sample tube in the heat block) and add the 150 µL aliquot to a tube with iodine solution

- mix the tube containing the starch solution, sample, and iodine. This represents the "end time" for the calculation of v_0.

- immediately record the time and the absorbance at 620 nm of the solution[d]

C. If the result corresponds to...

1. ...consumption of greater than 90% of the starch, then the activity of your sample is too high. Repeat the assay with one or more dilute sample (e.g., dilute 10-fold and 100-fold and reevaluate).

2. ...consumption of less than 5% of the starch, then consider repeating the assay (1) with a less dilute sample (although you will want to preserve the majority (avoid using more than 10% of the volume for this experiment) of your sample for analysis by SDS-PAGE) or (2) incubating the sample for a longer period of time (e.g., 10 min.) and/or at 37°C.

14.3.2. calculations

In general, enzyme activity can be expressed in International Units, defined as µmol substrate/(min * L). Since we are unable to measure the molar concentration of the substrate (starch) with this method, instead calculate activity using an arbitrary definition of 1 Unit of α-amylase activity as one mg/mL of starch lost per minute.

14.4. Laboratory report

This exercise is intended as only one component of a larger project which includes both the purification of α-amylase and characterization of the purification. The report should be prepared after the final laboratory session associated with the project. The report should be a comprehensive report, following the standard format, describing the materials, methods, and collected data from all lab sessions associated with the project.

15. Column chromatography

15.1. Introduction

In this lab, you will calibrate a gel filtration[1] column and use that data to estimate the molecular mass of salivary α-amylase using gel filtration chromatography. See "Estimation of the molecular weights of proteins by Sephadex gel-filtration" [2] for a historical perspective on the technique.

Before beginning the lab, take the time to review the meaning of the following terms:

> column chromatography
> gel filtration
> eluant
> void volume
> fraction

15.2. Procedure

15.2.1. Column preparation

At least four hours prior to column chromatography, prepare a column[2] and store it at 4°C.

---

[1] The 'gel filtration' method is sometimes referred to as 'molecular exclusion' or as 'size exclusion'.

[2] The course instructor may prepare the column in advance

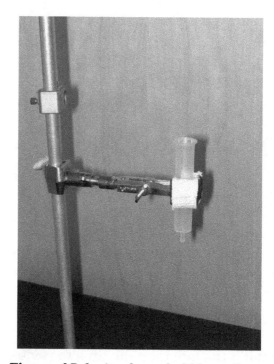

Figure 15.1. A column held by a clamp

1. Obtain the following:

   - column (PD-10)

   - column cover (or a piece of aluminum foil or plastic wrap)

   - Luer-lock stop-cock

   - pasteur pipet and a rubber bulb

   - fritted disc

2. Attach a clamp to a ring/support stand and attach the column to the ring.

3. Attach the stop-cock to the column.

4. Place the fritted disc in the base of the column.[3] Rinse

---

[3]If a fritted disc is not available, one can remove the stop-cock and pack

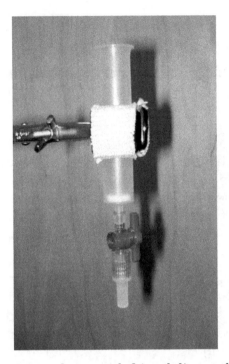

Figure 15.2. A column with fritted disc at the column base and with attached stop-cock

the column with the stop-cock in the open position. Move the stop-cock to the closed position.

5. Label the column near the bottom of the column with a piece of tape identifying the column as belonging to your group.

6. Close the stop-cock. Add approximately 1 mL of 0.1M NaCl into the empty column.

7. Shake the Sephadex G-75[4] slurry vigorously to evenly

---

the base of the column with a bit of cotton or glass wool, and replace the stop-cock on the column. However, this may result in a reduced flow rate.

[4]Sephadex G-75 is a size-exclusion (gel-filtration) chromatography medium composed of macroscopic beads consisting of cross-linked dextran; G-75 has an advertised fractionation range of 3000 to 80000

suspend the slurry. *Immediately* (before the slurry settles) proceed to the next step.

8. Fill the column with the slurry to approximately 5-10 mm from the top of the column. Cover the top of the column with the column cover.

> Be careful to avoid introducing bubbles into the column while adding the solid phase. If using a funnel to add the solid phase does not work well (this may be the case for small-volume columns), use a pasteur pipet (with bulb attached) to fill the column.

9. Store the column in the refrigerator in the vertical position (if necessary, use a beaker with paper stuffed in it to keep the column upright), allowing the solid phase to settle for at least two hours. Optionally, an additional 1 to 2 mL of eluant can be run through the column to encourage further settling.

15.2.2. Prepare the sample

> Although this lab is designed to be integrated into the larger 'purification and characterization of α-amylase' project, it can be conducted with any sample of interest provided an assay is developed to detect the compound of interest in the eluant.

1. Place approximately 1 mL of saliva in a 1.5 mL microcentrifuge tube. Centrifuge the tube at 8000 g[5] for 5 min.

2. Transfer 130 µL of the saliva supernatant to a new microcentrifuge tube. Add 40 µL blue dextran[6] dye solution.

Da.

[5] 9033 rpm if using a Fisher Scientific Marathon Micro A microcentrifuge

[6] A polysaccharide composed of glucose residues. Blue dextran behaves

Add 40 μL myoglobin[7] solution. Add 40 μL vitamin B12[8] solution. Mix well.

15.2.3. Column chromatography

1. Prepare tubes to collect the eluant (for starters, ensure you have at least 15 microcentrifuge tubes ready for eluant collection): number the tubes from 1 to 15.

2. Remove the column from the refrigerator. Place it in a stand on the bench in the vertical position. Use a marker to mark the exact location of the top of the bed.

3. Drain the liquid in the column until the level of the liquid is exactly at the top of the column bed. You may need to apply slight pressure to the column to initiate flow.[9]

4. Add 250 μL of the sample to the column.

> Be **very** careful when loading your sample to avoid disturbing the column bed: slowly "run the sample down the side of the column", distributing the sample across the top of the column bed as evenly as possible.

5. Open the stopcock and allow your sample to load onto the column. Close the stopcock as soon as the level of the liquid is *exactly* at the top of the column bed (don't allow the column to dry).

---

as an appr. 2×10^6 Da compound and absorbs strongly at 620 nm.

[7]Equine skeletal muscle myoglobin has a molecular mass of approximately 18800 Da and absorbs strongly at 555 nm.

[8]Vitamin B12 (cyanocobalamin) has a molecular mass of 1355.37 Da and absorbs strongly at 361 nm (but can be perceived visually as a light pink color).

[9]One can also aspirate the liquid at the top of the column to remove any suspended fines, avoiding disturbing the top of the column bed.

6. Add 400 µL of eluant (0.1 M NaCl), being very careful not to disturb the column bed. Open the stopcock and allow the eluant to move onto the column. Close the stopcock when the eluant is approximately 2 mm from the top of the column. Add an additional 200 µL of eluant, being very careful not to disturb the column bed.

7. *Carefully* add 2 mL of 0.1 M NaCl to the column, being careful to **avoid disturbing the column bed**.

8. Place the microcentrifuge tube number '1' under the column to collect the first fraction. Open the stopcock and fill the tube nearly completely. Close the stopcock.

9. Place the microcentrifuge tube numbered '2' under the column to collect the second fraction. Open the stopcock and allow eluant to drain out of the column, filling the tube nearly completely. Close the stopcock when the level of the liquid is approximately 2 mm above the top of the column bed.

10. Add an additional 3 mL of 0.1 M NaCl (carefully) to the top of the column.

11. Place the next tube, tube 3, under the column. Open the stopcock. Collect no more than 10 to 15 drops in this tube.

12. Continue collecting small fractions with tubes 4, 5, 6, ...:

    - Collect approximately 0.8 mL per *fraction* (this should correspond to approximately 10 to 15 drops).[10]

13. Stop when the eluant is approximately 1 mm from the top of the column.

14. Add additional eluant and collect additional fractions until at least 15 fractions have been collected.

---

[10]If the "fraction" concept is unclear to you, basically the idea is that, after collecting 0.8 mL, collect the subsequent 0.8 mL (the next fraction) in the next tube, and so on...

15. Analyze the eluant fractions (see below). Mix each fraction by vortexing before assaying.

16. Add additional eluant to the column and continue the chromatography if all analytes have not exited the column.

15.3. Analyze the eluant fractions

For each tube, perform an α-amylase analysis, a blue dextran assay, a vitamin B12 assay, and an equine myoglobin assay. Perform the non-destructive photometric assays first; then use the sample for an α-amylase assay.

15.4. Blue dextran, vitamin B12, and myoglobin assays

1. Blank a spectrophotometer using water in a clean cuvet.

2. Record the absorption values at 555 nm and 620 nm.

3. Remove the water from the cuvet and add the sample of interest.

4. Record the absorption values at 555 nm and 620 nm.

5. For vitamin B12, if a quartz cuvet and a UV-vis spectrophotometer are available, evaluate absorbance at 361 nm. Otherwise, visually assess the intensity of the light pink color of the tube, scoring each tube as ++, +, or 0.

6. Save the sample for use with the α-amylase activity assay.

15.5. α-amylase assay

Add 750 μL starch solution to a 1.5 mL microcentrifuge tube. Add 50 μL of the sample of interest to the starch solution. Immediately place the tube in the 37°C water bath. Incubate for 3 min.[11] Remove 150 μL of the sample and add to 600 μL iodine solution. Measure the absorbance at 620 nm.

15.6. Laboratory report

If this lab is conducted as part of the larger α-amylase purification and characterization project, the course instructor may indicate that the results from this exercise should be integrated into a larger report associated with the entire project. Ensure you determine whether a separate laboratory report should be prepared for this exercise.

Estimate molecular masses via size-exclusion column chromatography by first calculating K for each analyte where $K = \frac{V_e - V_o}{V_s}$ and

> V_e is the volume of solvent required to elute the compound of interest
> V_o is the void volume
> V_s is the volume of the stationary phase hint:[You will need to plan ahead for this; consider measuring the height and diameter of the column.]

- for most compounds, K and $\log M$ (M = molecular mass) are linearly related

- plot your data on a graph which allows you to estimate the molecular mass of α-amylase

---

[11]Alternatively, incubate at 25°C for 4 min.

- in a supplemental data section at the end of your report, show the raw data and calculations used in generating the linear fit which was used to estimate the molecular mass of α-amylase

16. SDS-PAGE

In advance of this lab session, you <u>must</u> prepare a "loading table" (see the "Planning" section below).

This experiment will require the full laboratory session time and, perhaps, additional time outside of the scheduled lab.

Polyacrylamide is a neurotoxin; wear gloves when handling polyacrylamide gels.

16.1. Introduction

Electrophoresis is a powerful means of separating molecules. Sodium dodecyl sulfate-polyacrylamide gel electrophoresis (SDS-PAGE) is frequently used to separate polypeptides on the basis of their molecular masses. In this technique, prior to electrophoresis, proteins are denatured by heating in the presence of the strong detergent, SDS. A reducing agent such as β-mercaptoethanol or dithiothreitol may be added to disrupt disulfide bonds. In this state, the number of SDS molecules bound to the polypeptide is approximately proportional to the mass of the protein. When subjected to an electric field in a polyacrylamide gel, these polypeptide-SDS complexes migrate with a velocity roughly proportional to their mass.

SDS-PAGE is typically used for protein analysis; variants of PAGE are also used for preparative work. SDS-PAGE can be used as a second dimension, in conjunction with isoelectric focusing, to effect a two-dimensional protein separation.

However, probably the most frequent use of SDS-PAGE is in conjunction with Western blotting: after subjecting a protein mixture to electrophoresis, the polypeptides in the gel are transferred to a membrane, either electrophoretically or via capillary action, and the membrane incubated with antibodies against a specific protein. Detection of the antibodies bound to the protein on the membrane serves as a highly specific and sensitive means of detecting a particular protein of interest, and of quantitating levels of that protein.

In this lab you will subject a sample containing (hopefully!) protein to SDS-PAGE. You will evaluate the results either by staining the SDS-PAGE gel or by Western blotting.

16.2. Procedure

16.2.1. planning

Before the lab session, plan how you will "load" the gel. "Loading" the gel refers to the process of adding a predetermined volume of each sample to a well in the gel. Each sample is added to a separate well (i.e., one well per sample). To load the gel, one must plan ahead and first determine the sample mass, and from that value, the corresponding sample volume, that will be loaded in each lane. Unless your instructor indicates otherwise, 10 µg of each sample should be loaded in a lane.[1] Prior to loading, each sample is heated in a solution of SDS and β-mercaptoethanol in order to completely denature the sample.

Use the information provided below along with the protein concentration estimates you obtained from the Bradford anal-

---

[1] If 10 µg of sample isn't available or will result in a loading volume greater than the capacity of the well, use the maximum amount of sample possible.

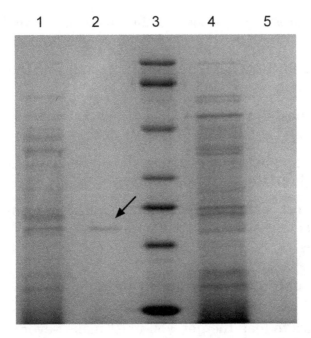

Figure 16.1. A SDS-PAGE gel. Lanes 1 and 4 represent electrophoresis of crude cellular extracts and lane 3 represents electrophoresis of a molecular mass protein standard mix (top to bottom: 90, 66, 45, 34, 27, 20, 14.4 kDa)

ysis of your unknown sample(s).

In addition, you will be provided with a pre-stained mass standard "ladder". For this sample, your instructor will indicate the volume of the sample which should be loaded. An example of a pre-stained molecular mass standard is Lonza's "ProSieve Color Protein Markers" mixture. This mass ladder containing polypeptides with apparent masses of 181, 121, 77, 48, 39, 25, 19, 12, and 10 kDa. The bands are prestained with different colors in order to aid the user in identifying the bands (the colors corresponding to the bands described above are purple, purple, red, red, purple, red, red, purple, and red, respectively).

Thinking ahead: some technical notes on SDS-PAGE

If you use SDS-PAGE in the future, you may find these point-
ers helpful in obtaining better results

- the best resolution is usually obtained with approximately 1 µg protein or less

- the sample loading volume should be less than 1/2 the
  height of the sample well

- for best results, load 1X sample buffer into any sample
  wells that are not loaded with sample

16.2.2. calculating loading values

To prepare to load the gel, you must make several calculations. For example, consider a gel in which you plan to load
10 µg of 'sample 23' in lane 8:

1. To add 10 µg of the sample, you must know the protein
   concentration of the sample (either determined by a protein assay you conducted or provided by the instructor).
   *If you already did a Bradford assay of an unknown, you
   should use your data for this calculation.* 'Sample 23' has
   an estimated concentration of 2.5 µg/µL. Given this, how
   many µL of sample correspond to 10 µg? This volume is
   the volume of sample you will place in the tube in which
   you prepare the final solution to be loaded onto the gel.
   This is also the value which you should enter in your loading table in the column labeled 'volume sample/µL'.

2. The sample should be adjusted to 1X sample buffer (1X
   SB) and the final volume of the sample to be loaded should
   be 15 µL. Therefore, the amount of 3X[2] sample buffer
   which should be added is 15 µL / 3. What is 15 µL / 3 ?

---

[2]"3X" implies a three-fold concentrate

lane	sample	μg sample	$\mu g/\mu L$ sample	μL sample	μL water	μL 3X SB	μL total
8	unknown	10	2.5	4	6	5	15

Figure 16.2. Example of a single row from a gel loading table

Figure 16.3. A screwcap tube and cap

This volume is the volume you should enter in your loading table in the column labeled 'volume 3X SB / µL'.

3. Calculate the additional volume of water required to bring the final volume to be loaded to 15 µL. If you have added 5 µL of 3X SB and 2 µL sample, how many more µL will you need to add to end up with a final volume of 15 µL? This amount should be entered in the column of the loading table labeled 'volume H_2O / µL'.

These calculations should be made for each gel lane to be loaded. (Just in case you didn't catch it, you should organize the calculated values into a table). See Figure 16.2.

16.2.3. preparing the samples

Prepare samples (unless directed otherwise, use a 15 µL loading volume)

1. Determine the sample volume, sample buffer volume, and amount of water that should be added to prepare the final sample to be loaded. If you skipped the guidelines given

above, now is the time to go back and read them more closely.

2. Obtain a 1.5 mL screw-cap tube (Figure 16.3) for each sample (screw-cap tubes are used to ensure that lids don't "pop open" during heating).

3. Directly before use, add β-mercaptoethanol (β-ME) to the 3X sample buffer solution. Prepare 3X sample buffer supplemented with β-ME by adding 5 µL of β-ME to 95 µL of 3X sample buffer. Use this modified 3X buffer in preparing your samples.

4. Prepare your samples in 1X SB (by adding the appropriate volumes of 3X buffer and water) using the loading table to guide you. Mix thoroughly. Keep samples on ice.

5. Denature your samples (prepared in 1X SB in the previous step) by heating for 5 min at 95°C in a heat block or in a water bath. Cool for 2 min on ice.

- Check with the instructor to determine whether you *should* or *should NOT* heat or denature the prestained mass marker sample (if such a sample is being used)
- It's probably not such a bright idea to immerse the sample tube in liquid during cooling (in at least one instance, students have discovered physics hard at work as liquid is 'sucked in' to the the tube as its contents cool).

16.2.4. preparing the gel electrophoresis apparatus

1. Rinse the gel electrophoresis box (Figure 16.4)

2. Obtain the SDS-PAGE gel (Figure 16.5). Note that, for many commercial precast gels, a piece of tape must be removed from the bottom or side before the gel can be used. Ensure your instructor has provided direction on this point before preceding.

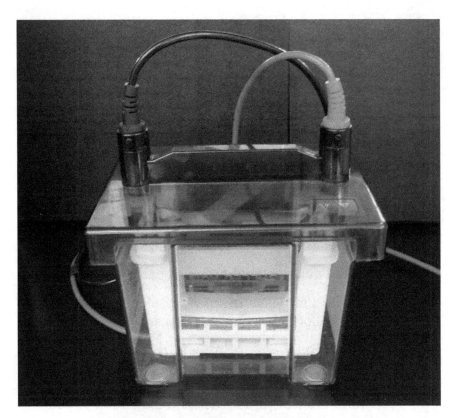

Figure 16.4. PAGE gel electrophoresis box

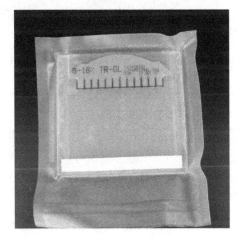

Figure 16.5. SDS-PAGE gel

3. Carefully remove the comb from the gel by sliding it with a slow and steady motion straight up and out from the cassette.

4. Rinse wells with 1X running buffer to remove unpolymerized or partially polymerized acrylamide fragments, etc. – this can be done (1) with a squirt bottle filled with running buffer or (2) with a syringe and needle. Be gentle, you don't want to damage the gel.

5. If the gel wells are not already outlined on the gel plate, use a permanent marker to draw lines underneath each well on the outside of the tall plate so that it will be easier to see well locations when loading the samples.

6. Assemble the gel electrophoresis apparatus:

   • Place the gel cassette into the electrode assembly with the short plate facing inward.

      note: For the MiniProtean III, make sure the gel is forced all the way down so the two blocks at the bottom hold the gel in place (this may take a bit of force – ask the instructor if you would like assistance).

   • Slide the entire assembly and cassette into the clamping frame.

      note: Make sure the rubber gasket (Figure 16.6) fits snugly against the gel - you may have to switch the gasket to the other side (the gasket should be placed so the flat side faces outward for Duramide gels) to get a tight seal.

   • Press down the electrode assembly while closing the two cam levers of the clamping frame.

   • Lower the inner chamber into the "mini tank".

   • Fill the inner/upper chamber of the apparatus with

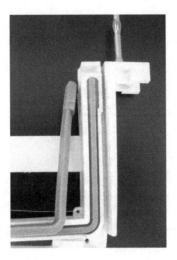

Figure 16.6. Gel gaskets

1X running buffer. Before proceeding, ensure that no leaking occurs.

16.2.5. loading samples and running the gel

1. For each sample:

    - Using a gel loading tip, withdraw sample from tube using a 20 μL pipettor *or* use a 10 or 20 μL syringe.

    - Ensure there are no bubbles or air in the sample.

    - Insert the tip into a well so that it is approximately 4 mm from the bottom of the well.

    - Slowly depress the plunger of the pipettor (so that the sample collects at the bottom of the well).

2. Carefully layer running buffer on top of the wells.

3. Ensure that the bottom electrophoresis chamber is filled with 1X running buffer (the upper and lower chambers together will require approximately 350 mL 1X running

buffer). If you have sufficient buffer (and if you are certain your apparatus is not leaking), fill the chamber to the bottom of the sample wells to keep the gel cooler.

4. Run the stacking gel at 100 V[3] until protein bands have passed from stacking gel into separating gel at least 1 cm.

5. Run the separating gel at 180 V (run until the dye front is approximately 1 cm from the bottom of the gel). This should take approximately 1 h.

6. Turn off the power supply and remove the top of the gel electrophoresis apparatus.

7. Remove the gel holder; remove the gel (and the two plates sandwiching it) from the holder.

8. For precast commercial gels, using a plastic or metal lever, twist the edge of the lever between the edges of the two plates so that the two plates separate.

9. Remove one plate slowly, allowing the gel to stick to the other plate.

At this point, ensure that you understand whether you will be staining the gel or performing a Western blot with the gel.

16.2.6. Colloidal Coomassie staining

note: your instructor may have you use conventional Coomassie staining or another staining technique instead of the procedure described below

---

[3]Depending on the type of gel and buffer being used, a higher voltage may be acceptable; check with your instructor to determine the appropriate voltage.

1. Fix the gel: immerse the gel for 60 min in 50% methanol/10% acetic acid.[4,5]

2. Obtain fresh stain solution.[6]

3. Immerse the gel in the staining solution. Seal the container. Rock gently for a minimum of 1-2 h (best sensitivity is with overnight staining - background rises after appr. 12 h staining).

4. Rinse with water or 1% acetic acid.

5. For a really clear background, destain in water several hours or overnight.

6. Store in 25% ammonium sulfate.

7. Briefly wash in 20% methanol (aq).

8. Record the results (a low-end scanner or camera works fine for this).

9. For stable storage (over 3 d), either

   1. store in 25% w/v ammonium sulfate rather than water and store at 4°C or

   2. rinse very thoroughly with water and place between cellophane sheets to dry.

---

[4]Fixing for a longer duration (e.g., overnight) is acceptable. The gel may shrink a bit but will expand when stained.

[5]One can also fix by incubating for 30 min w/gentle rocking in 12% trichloroacetic acid, 3.5% (w/v) 5-sulfosalicylic acid. However, TCA is a bit less student-friendly of a reagent.

[6]DIRECTLY before use, vortex the 5% dye solution (Coomassie Brilliant Blue G250 in water) and add 2 mL of the 5% dye solution to 98 mL of 15.6% ammonium sulfate, 0.29M phosphoric acid solution (15.6 g ammonium sulfate and 1300 µL of 85% phosphoric acid in 98 mL of water). Immediately mix this solution. Add 25 mL methanol and immediately mix the solution. This solution should not be stored but should be immediately used for staining.

16.3. Laboratory report

The laboratory report should follow the format described in the first laboratory session and in the course syllabus. This laboratory report should include data from both the previous laboratory session and this laboratory session.

Evaluate your data:

1. Characterize the nature of separation produced by SDS-PAGE: do your results support the assertion that the distance migrated is proportional to molecular mass? Is there a linear correlation between distance migrated and mass?

2. Evaluate the molecular mass(es) of the protein component(s) of the unknown sample by generating a calibration curve using the molecular mass standard(s) or the protein ladder. Note that here we are referring to a set of standards which correspond to proteins of known mass (e.g., 10 kDa, 30 kDa, 50 kDa, ...).

3. If you used mass loading standards (versus a conventional protein ladder), evaluate the concentration of the unknown protein by generating a mass calibration curve using the mass loading standards. Note that here we are referring to a set of standards which correspond to known masses of protein loaded (e.g., 0.1 µg, 1 µg, 5 µg, ...).

4. See also Appendix O.

16.4. Some resources

16.4.1. molecular mass determination

After staining molecular mass standards by a suitable method (if they are not prestained), a reference graph can be made by plotting the relative mobility, R_f,[7] of each standard versus the corresponding molecular mass. The relationship is approximately exponential: plotting R_f versus log(molecular mass) should yield a linear plot. The molecular mass of unknowns can then be estimated from the relationship described in the above plot.

16.4.2. sample concentration

Densitometry, the quantitation of optical density in a light-transmitting material, can be used to estimate the mass of protein present (i.e., how much protein was actually loaded in a given lane). This allows SDS-PAGE to function as a quantitative tool for estimating protein concentrations, providing data which parallels the estimates obtained using the Bradford assay. Some software packages with densitometry capabilities are listed in Appendix O.

---

[7]the distance traveled by the protein of interest divided by a reference distance (e.g., the distance traveled by a tracking dye)

17. Project reports for the α-amylase project

17.1. The laboratory report

The lab report for the α-amylase paper should be prepared on an individual basis (although you may work together in compiling and analyzing your data). The report is due one week after completing the last lab segment.

After the report is turned in, the reports will be redistributed to the class. Each class member will receive a report written by another class member. Your assignment for the upcoming week is to *carefully* review (evaluate and critique) the paper you receive.

17.2. Peer review of the α-amylase report

The report describing the laboratory exercises which dealt with purification and characterization of α-amylase is due one week after completion of the last experiment (SDS-PAGE of the various "α-amylase samples"). Your instructor will indicate whether the report should be prepared as a group or on an individual basis.

Whether the report is prepared in the context of a group or prepared individually, it is *your* responsibility to ensure you have on hand a copy of the submitted version of the report.

Each student or student group will receive one (or more papers) representing the reports submitted by another student or student group. These papers will be distributed (anonymously) at the course session following the session at which the laboratory reports were due.

A written review of the paper will be **due one week from the date on you received the paper(s) to be reviewed**.

Your instructor may also assign an oral presentation. If the reviews are prepared as groups, each member of the group is expected to participate in a significant component of the oral review presentation (i.e., if the review is conducted as a group, all group members are expected to speak...).

17.3. Background

At this point in the course, you have written multiple scientific papers (the lab reports you wrote earlier in the course) and may have read and evaluated one or more published peer-reviewed scientific articles. In this assignment you will review a scientific paper written by another member or members of the class and submit the review both as a written document and as an oral presentation to the class. Remember, for both components of the review, that criticism does not necessarily represent a personal attack on the authors of a paper. Instead, the role of the reviewer is to *assist* the authors of a paper in preparing a better paper. Every valid and constructive criticism that a reviewer raises with respect to a weakness of a paper represents an opportunity for the authors of the paper to improve their paper in a subsequent draft of the paper.

17.3.1. written review

Your review should be no more than **two pages** of single-spaced printed text. In an initial paragraph, the review should identify the primary hypothesis or hypotheses of the paper and briefly summarize the methodology employed to test the hypotheses. The review should then proceed to (1) summarize the data presented in the paper, and (2) indicate whether the data supports the hypothesis or hypotheses which the authors of the paper have articulated (this last part should receive special attention; the data in the report should be *thoroughly* evaluated in determining whether the conclusions drawn in the report are legitimate).

For each significant item in need of revision, the reviewer should describe the issue and precisely indicate the location of the problem by page and paragraph number or figure and table number. For example, it is not acceptable to simply indicate that "There were a few grammar errors" or "A few of the tables were missing units".

In a concluding paragraph, the review should provide an overview of the paper, identifying both the major shortcomings and primary strengths of the paper. Finally, the reviewer should recommend whether the paper be *accepted as is*, *accepted with revisions*, or *rejected* with a brief summary regarding the reasoning behind the recommendation.

If the reviewer recommends accepting the paper pending revision, the reviewer should clearly indicate the necessary revisions by listing and numbering them on a second sheet of paper. These can include revisions which require additional experimental work (even though, in this course, it won't be practical for the group to actually perform such additional experimentation).

If the reviewer recommends rejected the paper, the reviewer must justify the recommendation and by providing a brief, clear summary of the reasoning underlying the recommendation.

17.3.2. oral review

The oral presentation should use the structure suggested above for the written review as a suggested outline for organizing their presentation.

The presentation should not exceed the time limit indicated by the course instructor.

Consider using a computer or preparing overheads to facilitate ready discussion of the data in the paper and to supplement the oral component of the presentation.

Hints:

- be sure to focus on some positives during your presentation

- try and avoid focusing excessively on relatively minor issues such as formatting, grammar, and punctuation issues (while these should be noted, the *focus* should be on the results: what is the data? how well does the data support the hypotheses of the paper?)

- stand in front of your audience and look at them occasionally (vs. staring at a sheet of paper w/your notes for the entire presentation)

17.3.3. revisions

Each student or student group (depending on how the initial report(s) was/were submitted) will receive a copy of the written review of their paper and will have the opportunity to respond with a self-evaluation of their work (one page or less), accompanied by revisions if the manuscript reviewer suggested accepting the article with revisions or if the manuscript reviewer suggested rejection of the article.

Consider meeting with the course instructor (during office hours or by appointment) to discuss the project. You should bring in your written self-evaluation, the peer review, and your revised paper.

A. Sample lab safety handout

A.1. Etiquette

If you did not purchase it, were not given it, and you did not earn it, it probably does not belong to you.

Please do not remove laboratory items from the lab area without the instructor's permission.

Do not use/touch/move a sample that belongs to someone else (i.e., if a sample tube/vial/etc. does not have your initials on it).

A.2. Glassware and plasticware

Unless otherwise instructed by the course instructor, clean plasticware and glassware after use. The method used to clean labware depends on the contaminant.

>Hydrophobic (e.g. greases): Rinse with a dilute detergent solution (<2% strength) followed by thorough water rinsing. Alternatively, rinse with at least 70% ethanol or 70% isopropanol (avoid stronger solvents such as acetone or methanol with plasticware).

>Biological: Soak in a 1:10 dilution of bleach or use

wescodyne or 70% ethanol to disinfect; rinse extensively with water.

Hydrophilic (e.g., salts, buffers, etc.): Extensively rinse with water.

After cleaning, invert on a paper towel to dry.

A.3. Safety

Don't rush – relax – mistakes happen most easily when you are uptight or hurrying

In the case of an emergency, notify the instructor at once.

Note the locations of the

- emergency gas shutoff

- first-aid kit(s)

- fire extinguisher(s), fire alarm(s), and fire blankets

- eye-washes

- showers

Think out in advance how to deal with and prevent fires. Very small fires on the bench can be allowed to burn or smothered with wet towels. Dispose of flammable liquids in waste bottles in the hood. If a substance emits flammable vapors, keep it in the hood at all times – especially if burners are being used in the laboratory.

Pregnancy: if you are pregnant or planning on becoming pregnant, it is worth considering taking the class at another time since there is potential risk of exposure to teratogenic compounds.

Figure A.1. A safety shower and eyewash

A.3.1. The bench

Do not store "non-laboratory" items (e.g., your coat or text-books) on the bench – the goal is to keep chemical or micro-biological 'nasties' from migrating out of the lab with you

Clean the bench after use. Use the same method as would be appropriate for cleaning plasticware or glassware.

A.3.2. Using the hood

1. Make sure the hood is functional.

2. The hood sash should be closed when not in use.

3. The sash position where safe flow exists should be marked on the side of the hood. Closing the sash further increases the hood's effectiveness but may impair freedom of move-ment. It is generally a good idea when working with very volatile compounds to close the sash to the greatest ex-tent possible which still permits freedom of movement in conducting the manipulation. However, exercise care when working with powders or flames since lowering the sash can produce stronger air currents in the hood.

A.3.3. Apparel

Consider wearing safety glasses whenever you are doing laboratory work. In particular, wear safety glasses or gog-gles if there is a chance of splashing or an aerosol being formed – also wear safety glass during manipulation of vials during extreme temperature transitions. Contact lenses are NOT allowed.

Tie hair back – the goal is to keep your hair from wandering around in solutions, wrapping itself around lab equipment, or igniting in bunsen burners.

Wear your lab coat – this protects you from spills.

Wear close-toed shoes (i.e., no flip-flops, sandals, ...) – this protects you from spills.

In the case of a hazardous chemical spill, unless otherwise indicated, it is safest to immediately remove the clothing affected – if the chemical has reached the skin or is in the eye, rinse with copious amounts of water (locate the eyewashes in the laboratory).

Always wear gloves – make sure to choose the appropriate type of glove for the work you are doing. Latex gloves are appropriate for use with microbiologicals and water-soluble compounds. Nitrile gloves should be used when working with organic solvent-based solutions.

A.3.4. Personal hygiene

Wash hands and dry them after concluding an experiment and before leaving the laboratory – you probably don't want to eat chemical residues along with the food, gum, etc. you touch after leaving the lab.

Don't apply cosmetics or insert contact lenses in the laboratory.

Avoid eating, drinking, or tasting anything while in the laboratory. Avoid smelling lab reagents.

A.3.5. Accidents

Report accidental cuts/scrapes, burns, or chemical spills to the instructor immediately.

A.3.6. Liquid and solid manipulation

Information about laboratory hazards associated with a reagent is available in the form of a "Material Safety Data Sheet" (MSDS). The student should be familiar with the hazards presented by each compound which will be used in a given laboratory session. One means of acquainting oneself with the hazards presented by a compound is to review the MSDS for that compound. MSDS's are available online at several locations (e.g., `http://www.sigma-aldrich.com`).

Do not insert anything in a reagent bottle – including a "clean" spatula or dropper. Do not return excess material to a class reagent bottle. If a stopper or lid seems stuck, see the instructor.

Never mouth pipette – only use mechanical pipeting aids.

For powders which readily suspend (e.g., SDS), always use a face mask while measuring or pouring the powder.

If a substance produces fumes, it should be handled in a fume hood.

Spills: do not clean a chemical spill unless you know the proper means of neutralizing the spilled substance. If in doubt or unsure, notify the instructor. Acids and bases should be neutralized prior to any cleanup effort. Most other compounds should be adsorbed with activated charcoal.

Always add acid to water, never water to acid.

A.3.7. Working with flames

Don't – unauthorized flame use is not permitted in the lab. Flames present an extraordinary danger when used in the proximity of organic solvents, many of which are highly flammable.

B. Some writing and composition tips

B.1. Additional resources

For additional guidance in your writing, consider consulting one of the many texts on scientific writing. [1, 19, 6, 8, 14, 15, 17]

Several high-quality word-processing packages are freely available including LibreOffice (http://www.libreoffice.org/), OpenOffice (http://www.openoffice.org), and Abi-Word (http://www.abisource.com).

A number of packages are freely available for generating charts, graphs, and plots. The gnumeric (http://projects.gnome.org/gnumeric/), LibreOffice (http://www.libreoffice.org/), and OpenOffice (http://www.openoffice.org) spreadsheet components all have some charting capabilities. The stand-alone charting program gnuplot (http://www.gnuplot.info) is highly configurable, capable of two- and three-dimensional plotting, curve fitting (using an arbitrary function), and generating histograms. The R statistics package is also capable of generating high-quailty charts, graphs, and plots.

B.2. Self-Regulated Strategy Development Model

- leads to significant improvements in writing quality

Graham and Perrin 2006

> - concluded that SRSD has largest effect size of al writing interventions considered in a meta-analysis

Graham and Harris 2005

> - describes 20 validated strategies

Graham et al. 2000 'Self-regulated strategy development revisited...'

> - reviews evidence on its effectiveness

The examples and excerpts below are actual content from reports written by university undergraduates. If you seem to have 'special times' when writing reports and essays, consider also using a grammar checker.[1]

B.3. The art of the sentence

Ensure you understand the essential components of a well-constructed sentence.

B.4. The art of the paragraph

Use paragraphs to organize thoughts and to aid the reader in following the logic of the paper. If a single paragraph, single-spaced, takes up over 30% of a standard letter-sized page, it is time to ask yourself whether it would be a good choice to restructure the content into multiple paragraphs..

---

[1]There are several readily available resources such as http://afterthedeadline.com/ , http://www.grammarly.com/ , and http://www.languagetool.org/ .

B.5. Try using spellcheck functionality

Computers can help you. You might catch spelling errors.

example: "We then used a spectrohotometer to observe the trend..."

B.6. Learn to use the apostrophe

When referencing an item or object which belongs to another entity, you may be interested in the apostrophe character (').

excerpt: "The membrane was discarded and a classmates results were used as a comparison for this report."

Rewrite the above sentence correctly.

B.7. Learn to use, but not abuse, commas

excerpt: "Once, the molecules are subjected to electrophoresis the molecules are transferred to a membrane and subjected to Western blotting for detection."

As written, the above sentence is nonsensical. Rewrite it, changing only the position of the comma, in order to transform it into a sentence which is meaningful.

excerpt: "This portion directly relates to the western blot analysis (basically our final results) and so when done incorrectly results are inaccurate."

This sentence has multiple issues, including a need for commas to define phrasing. Rewrite the sentence so that it is grammatically correct.

example: "When observing the change of colors I became curious..."

Rewrite the sentence beginning, using the comma to better define sentence structure.

B.8. Learn to use the semi-colon

It is sometimes desirable to separate related sentences using a semi-colon character (;).

example of correct use of semi-colon: "The mobility of the protein in the gel is inversely proportional to its mass; the heaviest proteins move the slowest."

B.9. Adventure into the brave new world of avoiding repetitive sentence structures

Soporifics should be prescribed by a physician.

An excerpt from the 'Methods' section of a lab report: "We then decanted the buffer into a new 1.5 μL screw cap tube and heated it at 95°C for 5 minutes. We then used a Bradford assay to find the concentration of our protein samples. We used 9 different samples..."

What is the idea behind the phrase "stuck in a rut"? Do you think the style of writing exhibited above is helpful to the reader? Rewrite the above in a more concise and less repetitive style.

excerpt: "The sodium acetate was then also transferred into the beaker with the water and the glacial aceti acid. The solution was then stirred using the magnetic stir bar for approximately 3-5 minutes. The solution was then poured into a 100 mL graduated cylinder. De-ionized water was then added to the solution using a squirt bottle until the total volume was 100 mL."

Try rewriting this so that the sentence structures are a little more varied (consider addressing some of the other issues you may note as well...).

B.10. Try on conciseness for size. Avoid verbosity.

The scientific paper isn't a novel nor is it poetry. Unless one is very confident of the quality of his/her writing, it is typically best to suppress the urge to embellish.

example: "The sample was allowed to incubate at room temperature for 5 minutes."

better: "The sample was incubated at room temperature for 5 min."

example: "We saw the result from our experiment showing that bacteria species were present..."

This example has multiple deficiencies. One more concise rendering is: "Our experiment demonstrated that bacteria were present..."

example: "The sample was denatured by heating it in a water bath for five minutes at 95°C. The prestained mass marker sample was also heated for five minutes in the same bath."

Could the above excerpt be worded in a more concise manner?

B.11. Try and live in the now

excerpt from a the introduction of a submitted lab report:
"In these labs, we will try to extraction protein from ground
beef..."

Has the experiment been performed yet? If not, there is an
issue. If so, there is a different issue. Can you articulate the
problem associated with each possibility?

B.12. Learn to capitalize... and when
not to capitalize

1. Genus... capitalized. Species... lower-case. Genus...
   capitalized. Species... lower-case. Say it with me... Genus...

2. Capital letters are generally reserved for the start of a
   sentence or for a proper noun.

3. Contemplate the concept of a proper noun. Avoid the
   urge to transform regular nouns into proper nouns.

excerpt: "By using paper chromatography I tested my hypothesis: When autumn comes, the chlorophyll in the tree
leaves are completely lost..."

Rewrite this sentence so that it is correct (hint: problems
with capitalization are not the only issue challenging this
writer).

example: "If we were to repeat this experiment we probably
would need to measure the accuracy of the Spectrophotometer as the results we received were somewhat inconsistent
and we re-evaluated some samples multiple times."

Apart from being long and unwieldy, what should the case of
the first letter of "Spectrophotometer" be in this sentence?
Why?

example: "A Pediatric Endocrinologist treats children with complex hormonal disorders and other metabolic conditions."

Why are the words 'Pediatric' and 'Endocrinologist' capitalized? Should they be capitalized? If not, why not?

B.13. If in doubt, avoid a casual or personal tone

example: "By constructing a standard curve... one is able to determine the unknown concentration of their sample."

better: "The protein concentration of a sample can be calculated using a standard curve relating protein concentration to A_{590} / A_{450}."

example: All the cubes we placed into a 1.5 mL microcentrifuge tube."

Apart from being awkward, this sentence would be better off without the 'personal touch'. It's unclear whether the student meant "were" instead of "we" or whether the writing is simply poor. One thing is clear: the paper could have used another round of proofreading... Rewrite the sentence, improving structure and style.

excerpt: "Those are both of my hypotheses so why don't we get started with it."

While this might be suitable if the author were a talk show host, it's not the tone a scientific paper should adopt.

B.14. Who owns what? The possessive

example: "...is a normal part of the human bodies digestive tract."

example (after one round of instructor-requested revision): "...is a normal part of the human bodies' digestive tract."

Rewrite the sentence correctly. Explain why both of the student's attempts are not correct.

B.15. It's lonely to be alone... learn to avoid singular-plural dissonance in sentences

excerpt: "Since the hand sanitizers main ingredients is Ethyl alcohol, this is the most likely reason they killed the most bacteria..."

Rewrite the sentence correctly (hint: problems with singular-plural dissonance are not the only issue facing this author).

B.16. Eschew vagueness and ambiguity

example: "Dairy products contain an expiration date clearly indicated on the label that makes it susceptible to spoilage."

What is the problem with this sentence?

example: "We mixed all three materials together and placed it in a microwave for two minutes."

What is the problem with this sentence?

B.17. Exert effort.

Instead of "1x10^-7"... consider whether it is really that difficult to summon up the effort to find the 'superscript' functionality on the word processing software being used.

Apply the same principle to other formatting. For example, the 'a' should be subscripted in pK_a and the 'eq' should be subscripted in K_{eq}.

B.18. Hyphenate when appropriate

Can you identify ways in which the excerpts below might be improved?

"The most common effect on humans is diarrhea, which is not usually life threatening in developed areas."

"... who had suffered greatly from poliomyelitis at a young age and became ventilator dependent as the years passed."

B.19. Use abbreviations correctly

Except for common abbreviations, an abbreviation should be spelled out the first time it is used in the text (e.g., "...200 L of DMSO (dimethylsulfoxide) were transferred to the tank. The DMSO was then boiled.").

B.20. Adhere to scientific conventions with numeric values and units

When referring to a value, ensure that the numeric value is accompanied by a unit of measure, whenever appropriate. Furthermore, ensure that there is a space between the numeric value and the unit of measure.[2] For example, "200

---

[2]With some units of measure it is considered acceptable to omit the space is often not included. These include units of degrees (e.g., 30°C) and units of molarity or normality (e.g., a 1N HCl solution)).

mL of water" is correct. "450mm from the edge of the gel" is incorrect.

When referring to a decimal value with an absolute value less than one, reach deep inside and summon up the effort to include the zero character.

example: ".081578g"

Identify the errors associated with the above attempt to refer to a mass.

B.21. Don't plagiarize

Unless it's common knowledge – or a fact which you established first-hand – cite the source.

example: "...As the purification process progreses, some activity in evitably lost. That is unless an inhibitor is removed activating alpha-amylase. One of the more reknowned inhibitors of alpha-amylase is Streptomyces tendae. The more purified the sample becomes the higher the purification level..."

The above excerpt exhibits a number of deficiencies. However, one is much more severe than the others – can you identify it?

C. Spreadsheet software

C.1. Spreadsheet applications

A number of papers raise concerns regarding the errors in the proprietary "Excel" spreadsheet and suggest that other packages be used instead. [21, 28] The gnumeric (http://www.gnumeric.org/) spreadsheet application includes charting capabilities, statistical functions, and more. McCullough recommends using gnumeric instead of Excel if errors in statistical functions are a concern. [20]

For those most comfortable with something similar to "Excel", the spreadsheet program included in the LibreOffice office suite[1] may be the most straightforward to use.

A number of spreadsheets are available for GNU Emacs[2], including 'dismal' and 'ses'. The emacs wiki (http://www.emacswiki.org) has, in the past, been a good point of reference for these pieces of software.

---

[1]http://www.libreoffice.org
[2]http://www.gnu.org/software/emacs/

D. Statistics software

D.1. Online statistics resources

A wide variety of statistical tools are accessible online.[1] These resources range from relatively powerful tools such as Rweb,[2] a web-based interface to the R statistical package, to SISA,[3] a straightforward web-based interface to tests such as the T-test and calculations such as Bonferroni correction, to the 'Simple regression page',[4] a web-based interface for performing linear regression.

D.2. Statistics software

Most spreadsheet software (see Appendix C) includes some statistics functionality. Take the time to locate and use the documentation for the software package you use.

gfit Model-based global regression. Accessible at `http://gfit.sourceforge.net`.

gnuplot Linear regression and more sophisticated fits. Accessible at `http://www.gnuplot.info`.

---

[1] In the past, Interactive Statistical Calculation Pages (`http://statpages.org/`) has featured a relatively comprehensive list of web pages that perform statistical calculations.

[2] `https://rweb.stat.umn.edu/Rweb/`

[3] `http://www.quantitativeskills.com/sisa/index.htm`

[4] `https://www.zweigmedia.com/RealWorld/newgraph/regressionframes.html`

PSPP Aims to replace functionality of SPSS statistics package (including descriptive statistics, T-tests, linear regression and non-parametric tests). Accessible at `http://www.gnu.org/software/pspp/pspp.html`.

R A statistics package with a lot of power but also with a substantial learning curve; consider using one of the R GUIs to get started. Accessible at `http://www.r-project.org`.

E. Using the pH meter

Specifics of pH meter use vary a bit from instrument to instrument. Given this, the course instructor may provide you with instructions specific to the make/model if pH meter you're using. However, irrespective of the make and model of the meter, you should follow a few important guidelines:

1. treat the pH electrode with care (the electrode is typically relatively fragile)

2. rinse the pH electrode with deionized water before and after use

3. when finished with the pH electrode, return it to the storage solution

E.1. Taking a pH measurement

1. Before calibrating or taking a measurement, the pH electrode should be prepared for use. If the electrode isn't a sealed electrode, the electrode may have a small rubber plug inserted in a hole near the top of the electrode. This plug should be removed prior to taking a measurement (to facilitate even buffer flow across the electrode junction).

2. Before making a measurement, the instrument should be calibrated with a two- or three-point calibration. The specifics of calibration vary from instrument to instrument. Calibration not only ensures an accurate measurement but also serves as an opportunity to check the per-

formance of the pH electrode by evaluating response time and slope value.[1]

3. Once the instrument is calibrated, take the pH measurement:

   - rinse the electrode

   - immerse the electrode in the sample solution

   - remove the electrode from the sample solution

   - rinse the electrode

   - returning the electrode to the electrode storage solution.

E.2. References

The February 2004 issue of American Laboratory has a relatively readable article, "Improved pH Measurement" (by Piero Franco, Detmar Finke, and Peter Hail), describing pH measurement.

---

[1]pH electrodes, even well-maintained, will eventually reach the end of their useful lives and require replacement (a pH electrode normally lasts 6 months to a year under heavy use; the lifespan is shortened if the electrode is exposed to highly aggressive or strongly acidic samples). Near the end of the electrode's life, response time will deteriorate.

F. Graphing/plotting software

Many different applications are available for generating graphs, charts, or plots. Each application has its own strengths and weaknesses.

gnuplot, fityk, and SciDAVis provide support for **fitting data**.

For **plotting a function**, gnuplot and kmplot are recommended. For **plotting data**, gnuplot or SciDAVis is recommended.

F.1. software descriptions

fityk (http://fityk.nieto.pl/) Fityk includes the capacity for differentiation over a curve. The Macintosh and Windows versions of the software are not free.

gnumeric (http://www.gnome.org/gnumeric) The spreadsheet program gnumeric supports plot types including line, x-y, scatterplot, bubble, pie, area, bar, column, radar, ring.

gnuplot (http://www.gnuplot.info) Gnuplot is a 'tried and true' utility for generating charts. Most of the gnuplot GUIs are pretty klunky, though (easiest to use from command-line). Gnuplot's features include two- and three-dimensional plotting, curve fitting (using an arbitrary function), and a histogram tool.

openoffice.org (http://www.openoffice.org) The openoffice.org office suite includes a spreadsheet program which supports several plot types.

origin (http://www.originlab.com) Origin is commercial (non-free) software which is both easy to use and relatively powerful.

SciDAVis (http://scidavis.sourceforge.net/) SciDAVis is both easy to use and relatively powerful.

G. Molecular visualization tools

For viewing large biomolecules, both pymol and VMD are usable and featureful. When it comes to generating and optimizing small-molecule structures, ghemical is about the only option with a straightforward user interface.

G.1. Molecular structure viewers and utilities

ACD/3D viewer Good for small molecule viewing. From Advanced Chemistry Development. An 'add-on' to ChemSketch.

Avogadro A small molecule editor and visualizer. Historically, Avogadro has been available for Windows, Linux, and macOS. The user interface is suitable for undergraduate student use. Accessible at `https://avogadro.cc`.

bkchem A free tool useful for generating quick two-dimensional small molecule 'skeleton/zig-zag' structures. bkchem is capable of exporting in SVG, CML, or CML2.

ccp4 The ccp4 package includes 'ccord_format' (fix corrupted pdb files, etc. and convert to other formats), 'distang' (list all close contacts (distance/angle calculation)), 'hgen' (generate hydrogen positions for proteins), 'pdbcur' (pdb manipulations – can generate chains, rename chains,

...), 'pdbset' (various pdb manipulations), and 'water-tidy' (generate useful water naming system). Accessible at `http://www.ccp4.ac.uk`.

chimera Chimera has a nice GUI interface. It is a relatively easy to use utility for visualizing structures. Chimera can import and read PDB files. Chimera doesn't support any sort of 'molecule editing'. Accessible at `http://www.cgl.ucsf.edu/chimera`.

Cn3D Cn3D is a helper application for your web browser that allows you to viewing biomolecule structures. Very nice three-dimensional representations. One downside to Cn3D is that it does not have a print feature. Use Alt-PrtScr and then paste into a print-capable application to work around this little inconvenience. A second downside is that it uses ASN-1 - formatted files. Accessible at `https://www.ncbi.nlm.nih.gov/Structure/CN3D/cn3d.shtml`.

ghemical Ghemical is a nice computational chemistry package including an editor for creating molecules and the capacity to import/export PDB, MDL MOL, CML, ... (relies on openbabel). Accessible at `http://www.bioinformatics.org/ghemical/ghemical/index.html`.

gopenmol gopenmol has a nice GUI interface and is straightforward to install. gopenmol imports CML, PDB, ... and can perform some simple molecule editing operations (break/make bonds actions). Accessible at `https://research.csc.fi/-/gopenmol_download`.

moplot Moplot displays molecular orbitals, geometries, etc. Accessible at `http://moplot.sourceforge.net`.

openbabel Openbabel facilitates interconversion between file formats as well as providing various other functions. Accessible at `http://openbabel.org/wiki/Main_Page`.

pymol Pymol is a molecular visualization package which includes the capacity to rotate around a single bond. Pymol has historically been available for Linux as well as for Windows and macOS. Accessible at `http://pymol.org`.

QuteMol QuteMol is a molecular visualization package which accepts standard PDB input. Accessible at `http://qutemol.sourceforge.net`.

reduce Utility which can add hydrogens and optimize structure. The default mode optimizes one side-chain at a time. 'build' optimizes H-bonding networks.

DeepView swiss-pdb viewer (spdv) Spdbv has some nice features such as the capacity to quickly visualize hydrogen bonding patterns. Accessible at `https://spdbv.vital-it.ch/`.

UGENE UGENE is free open-source cross-platform bioinformatics software which includes everything from a molecular structure viewer which handles PDB input to sequence analysis tools ranging from sequence alignment tools to a restriction enzyme finder tool. Accessible at `http://ugene.net`.

VMD VMD facilitates visualization and manipulation of biomolecular structures. VMD can interact with namd and imd to do molecular dynamics simulations. VMD is not released as a debian package. Accessible at `http://www.ks.uiuc.edu/Research/vmd`.

YASARA View YASARA View is available for free. This cross-platform molecular visualization platform handles PDB files and has a host of other functions available as well. Accessible at `http://www.yasara.org/products.htm`.

H. Computational tools for protein structure

H.1. protein sequence database(s)

NCBI Protein A large variety of sequence data from many organisms and sources. The database can be searched (using Boolean operators) using many search fields – e.g., Organism [ORGN], Title Word [TITL], etc. For example, a search could be 'hmp[TITL] AND "Escherichia coli"[ORGN]'. Accessible at `https://www.ncbi.nlm.nih.gov/protein/`.

Human protein reference database A human proteome database including domain structure, post-translational modifications, interaction networks, ... Accessible at `http://www.hprd.org`.

H.2. protein homology analysis

- see Appendix I

H.3. protein structure viewing utilities

- see Appendix G

H.4. secondary structure prediction

gor iv See also NPS@. Accessible at `http://npsa-pbil.ibcp.fr`.

nnpredict See also NPS@. Typically, a number of web interfaces to nnpredict are accessible.

NPS@ NPS@ is an online software tool for protein sequence analysis hosted by Pole Bio-Informatique, Lyonnais. NPS@ includes access to over ten different protein secondary structure prediction utilities, including GOR IV, PHD, and PREDATOR. Accessible at `http://npsa-pbil.ibcp.fr`.

peptstats Calculates a number of statistics based on the primary sequence of a protein. These statistics include molecular mass, charge, isoelectric point, and molar extinction coefficient. pepstats has been accessible at `http://emboss.bioinformatics.nl/cgi-bin/emboss/pepstats` and `https://www.ebi.ac.uk/Tools/seqstats/emboss_pepstats/`.

PREDATOR The PREDATOR protein secondary structure prediction utility requires a sequence in FASTA format with no spaces in the amino acid sequence and including a >proteinname line. PREDATOR has been accessible at `http://www.embl-heidelberg.de/argos/predator/run_predator.html` and `http://bioweb.pasteur.fr/seqanal/interfaces/predator-simple.html`.

PredictProtein PredictProtein is a utility which predicts several aspects of protein structure; takes several minutes to generate prediction. Not the most user-friendly output. Accessible at `https://www.predictprotein.org/`.

prof PROF is a secondary structure prediction system. Accessible at `http://www.aber.ac.uk/~phiwww/prof`.

psa The Protein Sequence Analysis (PSA) predicts secondary structure and other folding classes. Accessible at `http://bmerc-www.bu.edu/psa/request.htm`.

PSIPred PSIPred can be used to predict several aspects of protein structure. The "Predict secondary structure" option will return an e-mail (after several minutes) describing the predicted regions of the protein with random coil, helical, or sheet secondary structures. A link at the bottom of the e-mail allows the user to see a graphical representation of the results. The "Predict Transmembrane Topology" option uses the MEMSAT program and returns an e-mail (typically fairly quickly) describing predicted transmembrane segments of the protein. PSIPred has been accessible at `http://insulin.brunel.ac.uk/psipred` and `http://bioinf4.cs.ucl.ac.uk:3000/psipred`.

simpa SIMPA is a method for predicting protein secondary structures. Accessible at `http://bmerc-www.bu.edu/psa/request.htm` and `http://abs.cit.nih.gov/simpa/index.html`.

H.5. domain/fold recognition/prediction

bioinfobank Bioinfobank is a "meta server" offering a number of different services including profsec, psipred, esypred3d, ffas03, grdb, pfam-basic, pfam-metabasic, and others. The service has been accessible at `http://meta.bioinfo.pl/submit_wizard.pl`.

inub Fold recognition. Results are sent via email. Minimum sequence length of 40 residues. Maximum sequence length of 990 characters. Accessible at `http://inub.cse.buffalo.edu`.

CDART CDART is a protein domain analysis utility which identifies domains present in a linear amino acid sequence. Results are returned in the context of a user-friendly web page. Analysis of longer sequences is permitted. CDART has been accessible at `http://www.ncbi.nlm.nih.gov/Structure/lexington/lexington.cgi`.

CDM This server accepts sequences up to 1000 amino acids. Accessible at `http://gor.bb.iastate.edu/cdm`.

dompred This server can require a few minutes before returning results. In the past, the structure of the results page has had some imperfections, including broken links. Accessible at `http://bioinf.cs.ucl.ac.uk`.

ffas ffas only accepts sequences less than 1000 aa. Accessible at `http://ffas.ljcrf.edu/ffas-cgi/cgi/ffas.pl`.

FOLD server This server uses a multi-stage procedure including BLAST, PSI-BLAST, and/or PSI-PRED. Accessible at `http://www.doe-mbi.ucla.edu/Services/FOLD`.

phyre phyre (Protein Homology/analogY Recognition Engine) is a protein fold recognition server. In 2008, the server had a 1200 residue limit. The prediction is based on protein fold recognition; phyre uses a fold-recognition system designed to model the entropy of a folding protein; additional modules from the Phragment ab-initio package rebuild protein loops and refine the final model using a combination of energy functions and Monte Carlo sampling. In 2010, analysis of a short oligopeptide took less than 30 min. Phyre can output results of matching folds in PDB format. The service has been accessible at `http://www.sbg.bio.ic.ac.uk/~phyre`.

SMART SMART (simple modular architecture research tool) identifies tertiary structure elements (does not generate a complete tertiary structure prediction based on a primary sequence). SMART accepts longer sequences.

Accessible at `http://smart.embl-heidelberg.de`.

H.6. tertiary structure prediction

3djigsaw 3djigsaw uses a homology modelling approach. Other sites which have offered 3djigsaw functionality include `http://bmm.cancerresearchuk.org/~3djigsaw/`. With version 3.0, use 'automatic' mode unless there is a specific reason to use a different mode. Results are returned via email. In 2010, a prediction job for a short oligopeptide was completed in less than 4 h. Accessible at `http://www.bmm.icnet.uk/servers/3djigsaw`.

CPHmodels CPHmodels uses homology modelling to predict tertiary structure. Results can be returned as e-mail or via the web interface. In the past, CHPmodels has exhibited difficulty if the sequence isn't similar to one of the templates in the database. CPHmodels has been accessible at `http://www.cbs.dtu.dk/services/CPHmodels`.

esypred3d An espred3d prediction job can take in excess of 2 days. The utility can output results in PDB format. The utility uses a homology modelling approach. The service has been accessible at `http://www.fundp.ac.be/urbm/bioinfo/esypred`.

I-TASSER Online platform for tertiary structure prediction. The service has been accessible at `http://zhanglab.ccmb.med.umich.edu/I-TASSER`.

SWISS-MODEL The SWISS-MODEL server facilitates several types of structure prediction. Results are sent by e-mail. In the past, the web interface required creation of a 'workspace' associated with an email address. Consider reading some of the documentation to learn how to use the SWISS-MODEL web interface. The 'fully automatic mode' (Modelling → Automated Mode) requires only the amino acid sequence - the software automatically selects a suitable template. The 'model coordinate' file can be exported as PDB. In 2010, a prediction job for a short oligopeptide was completed in less than 4 h. The service has been accessible at `http://swissmodel.expasy.org`. See also `http://www.expasy.ch/swissmod`.

SAM-T08 SAM-T08 outputs results in CASP 'TS' format (3D atomic coordinates) - submit as 'PDB' at `http://polyview.cchmc.org/`. SAM-T08 is advertised as providing several sets of results in addition to a predicted three-dimensional structure. However, it does not reliably do so. In 2010, a short oligopeptide was evaluated in less than 4 h. The software is described in "SAM-T08, HMM-based protein structure prediction" (Karuplus 2009). The author has indicated that, as no funding has been forthcoming for the web service, the service will become unavailable when the hardware it runs on ceases to function. The service has been accessible at `http://compbio.soe.ucsc.edu/SAM_T08/T08-query.html`.

H.6.1. ab initio modelling

PROTINFO AB CM Protinfo predicts tertiary structure using RAMP software (the 'de novo' option gives ab initio prediction (probably best for polypeptides with under 100 residues). This can take two to three weeks to complete. Accessible at `http://protinfo.compbio.washington.edu/protinfo_abcmfr`.

Robetta Robetta allows both ab initio and homology modelling predictions of tertiary structure. Note that ab initio prediction is very computing power-intensive. Allow a minimum of 72 h to obtain a complete ab initio prediction. In the past, an academic email has been required in order to register to use Robetta. Accessible at `http://robetta.bakerlab.org`.

hmmstr ROBETTA should provide a higher-quality prediction relative to that generated by the hmmstr server. hmmstr predicts the structure (secondary, local, supersecondary, and tertiary) of proteins from the primary sequence using a hidden Markov model approach. The current interface also includes the ability to output PDB via Rosetta. The hmmstr server was not accessible on 2008-02-12 nor on 2010-02-16. Accessible at `http://www.bioinfo.rpi.edu/bystrc/hmmstr/server.php`.

H.7. protein structure databases

NCBI Structure An interface to a database of protein three-dimensional structures. Accessible at `https://www.ncbi.nlm.nih.gov/structure`.

Protein Data Bank This is the primary database of protein three-dimensional structures. Accessible at `http://www.rcsb.org/pdb`.

H.8. miscellaneous utilities

ProtParam - predict physico-chemical properties of a polypeptide sequence (amino-acid and atomic compositions, pI, extinction coefficient, etc.). Access at `http://www.expasy.ch/tools/protparam.html` or `http://www.expasy.org/tools/protparam.html`.

Compute pI/Mw local This tool predicts the pI and molecular mass of a polypeptide based on primary sequence. Accessible at `http://www.expasy.org/tools/pi_tool.html`.

ScanSite pI/Mw The ScanSite pI/Mw page is an interface for predicting the pI, molecular mass, and identifying potential phosphorylation sites based on primary sequence. Accessible at `http://scansite.mit.edu/calc_mw_pi.html`.

proteine If you can speak French... Proteine predicts pI, etc. and calculates and displays a titration curve. Accessible at `http://www.iut-arles.up.univ-mrs.fr/w3bb/d_abim/compo-p.html`.

H.8.1. helical wheels

J. Everett's 'Helical Wheel program' `http://www-nmr.cabm.rutgers.edu/bioinformatics/Proteomic_tools/Helical_wheel/`

heliquest - includes functionality for analyzing a sequence and displaying hydrophobicity, hydrophobic moment, helical wheel, and other functions of the sequence. Accessible at `http://heliquest.ipmc.cnrs.fr`.

M. Turcotte's HelixWheel Java applet `http://biobug.life.nthu.edu.tw/predictprotein/tools/helicalWheel/`

PepWheel (EMBOSS) PepWheel displays a helical representation of a peptide sequence (i.e., "looking down" the axis of the helix). Many of the many EMBOSS webfrontends have included access to PepWheel.

I. Sequence analysis tools

I.1. base calling

e-Seq (Li-Cor) A commercial base calling package for Li-Cor sequencers.

phred A command-line utility to read sequencing trace information in the SCF, ABI, and ESD formats.

I.2. software suites

EMBOSS (http://emboss.sourceforge.net) The EMBO software suite is a collection of molecular biology applications. The EMBOSS software can be installed on a local computer (e.g., biolinux.org maintains fedora rpms EMBOSS and EMBOSS-Jemboss). GUIs and web interfaces to the EMBOSS tools are also available.[1] EMBOSS is a part of several live CDs including bioknoppix and vigyaan (vigyaancd.org).

FASTA The FASTA software suite is a collection of molecular biology applications (e.g., lalign). Many of these applications can be accessed via the web; the FASTA software can also (in theory) be installed on a local computer. However, fasta2 (the package with lalign) has not always been straightforward to build and install. GUIs for the FAST tools include visual FASTA and

---

[1]See http://emboss.sourceforge.net/interfaces.

its successor, octopus (note: these are no longer maintained).

Geneious (http://www.geneious.com/) This commercial software package is offered with a 14-day free trial for 'pro' features. The relatively usable gui for facilitates alignment, tree, assembly, and other features. Adding a new sequence only possible with the 'pro' features.

I.3. sequence database(s)

NCBI (https://www.ncbi.nlm.nih.gov/) NCBI provides a number of interfaces to access sequence data including Genome (https://www.ncbi.nlm.nih.gov/genome), GenBank (https://www.ncbi.nlm.nih.gov/genbank/), and NCBI Protein (https://www.ncbi.nlm.nih.gov/protein/). The databases have a large variety of sequence data from many organisms and sources. The nucleotide database includes sequence data including genomic, mRNA, and vector (e.g., plasmid) sequences from a wide variety of organisms. Use of Boolean operators (AND/OR/NOT) and search fields (e.g., Gene name [GENE], Journal name [JOUR], Organism [ORGN], Protein name [PROT], Title word [TITLE]) allows for powerful and efficient searches.

Promega (http://www.promega.com/vectors) This site contains sequences and maps for many of the commonly used vectors.

I.4. nucleotide sequence alignment

I.4.1. local alignment

lalign William Pearson's lalign program is part of the FASTA package. The lalign program uses the algorithm of Huang and Miller ref:[Huang91] and is designed to compare two protein sequences. The software has been accessible online at `http://fasta.bioch.virginia.edu/fasta/lalign2.htm`, `http://www.ch.embnet.org/software/LALIGN_form.html`, and `http://www.isrec.isb-sib.ch/experiment/ALIGN_form.html`.

LFASTA LFASTA is designed to work with nucleotide sequences. LFASTA can be accessed online at `http://pbil.univ-lyon1.fr/lfasta.php`, `http://www2.igh.cnrs.fr/fasta/lfasta-query.html` .

EMBOSS The EMBOSS program suite includes 'water' (Smith-Waterman local alignment) and 'matcher' (local alignment).

I.4.2. global alignment

ALIGN Part of FASTA package, ALIGN is designed to perform global alignments.

CLUSTALW CLUSTALW, a tool designed for *nucleotide* sequence alignment, has been accessible at `http://dot.imgen.bcm.tmc.edu:9331/multi-align/Options/clustalw.html` and `http://www.ebi.ac.uk/clustalw`.

needle Needle is an EMBOSS program for performing Needleman-Wunsch global alignment. Accessible at `http://analysis.molbiol.ox.ac.uk/pise_html/needle.html`.

stretcher Stretcher is an EMBOSS program for performing global alignment of two sequences. In the past, Stretcher has been accessible at `http://analysis.molbiol.ox.ac.uk/pise_html/stretcher.html`.

I.4.3. multiple sequence alignment

dialign The dialign utility can output sequence tree data if option checked. In the past, the dialign software has not been freely downloadable but several web interfaces have been available.

clustalw The clustalw utility can perform multiple alignments of protein or DNA sequences. The shell interface is relatively easy to use, typically generating an output file 'foo.aln' representing the sequence alignment. Online interfaces have been available at http://www.ebi.ac.uk/clustalw and http://bioweb.pasteur.fr/seqanal/interfaces/clustalw.html.

emma Emma is an EMBOSS utility capable of performing multiple alignments. Tools include infoalign (provides info on a multiple alignment), showalign (shows a multiple sequence alignment), and cons (calculates a consensus from multiple alignments).

I.4.4. visualization of multiple sequence alignment data (phylogenetic trees, etc.)

phylip Phylip facilitates generation and analysis of phylogenetic trees. The software seems to be more comprehensive than clustalx or jalview. Executables are available for windows, macintosh, and linux.[2]

phylodendron Phylodendron facilitates online visualization of phylogenetic data (the input format is phylip; the output is as a pdf representation of the tree). Accessible at http://iubio.bio.indiana.edu/treeapp/.

treeview Treeview can be used for visualization of phylogenetic data. Versions have been available for mac, win-

---

[2]http://evolution.genetics.washington.edu/phylip.html

dows, and linux. The software is both easy to install and use.

I.5. restriction enzyme resources and plasmid tools

New England Biolabs The restriction enzyme company.[3] Online catalogs, technical resources, vector sequences and maps, etc.

NEBCutter This online tool[4] evaluates restriction enzyme sites in a sequence of interest. Nice graphical representations, etc. Remember to indicate if the sequence is circular or linear.

ReBase This online database[5] of restriction enzyme characteristics contains links to a number of tools for identification of restriction sites in your molecule of interest. REBsites (`http://tools.neb.com/REBsites/index.php3`) is an online tool which gives convenient graphical representations of restriction digests of a sequence of interest. Remember to indicate if the sequence is circular or linear.

WebCutter This online tool[6] evaluates restriction enzyme sites in a sequence of interest. Remember to indicate if the sequence is circular or linear (select "Circular Sequence Analysis").

BioTools (`http://biotools.umassmed.edu/`) BioTools@UMass Medical School is an interface to utilities including online analysis of restriction enzyme sites in a sequence of interest.

---

[3]`http://www.neb.com`
[4]`http://tools.neb.com/NEBcutter/index.php3`
[5]`http://rebase.neb.com/rebase/rebase.html`
[6]`http://www.firstmarket.com/cutter/cut2.html`

I.6. protein homology analysis

I.6.1. binary comparisons

DOTLET A fairly easy-to-use online tool for aligning two protein sequences. The utility gives results in several different formats. Java-based. Difficult to cut and paste data from the applet. Accessible at `http://www.isrec.isb-sib.ch/java/dotlet/Dotlet.html`.

lalign See the nucleotide sequence analysis section above.

SIM This online tool[7] for binary comparison of protein sequences is straightforward to use. The default alignment parameters are sufficient for typical applications. Easy to cut and paste text alignment result into other applications.

I.6.2. homology search across a database

BLAST (`http://www.ncbi.nlm.nih.gov/BLAST`) BLAST has historically been the standard for searching for nucleotide sequences homologous to a sequence of interest. The tool is relatively user-friendly. Standard nt-nt BLAST is recommended for general homology searches. BLAST can also be used to conduct protein-protein searches for homologous sequences in the polypeptide sequence database. The search can modified in a number of fashions (e.g., by limiting the search to specific organisms or organism groups).

I.6.3. multiple comparisons

clustalw See the nucleotide sequence analysis section above.

---

[7]`http://www.expasy.ch/tools/sim-prot.html`

158

MUSCA Multiple alignments of sequences (sequences must be submitted in FASTA format). Not extremely user-friendly. Note that K=number of sequences. The service has been accessible at `http://cbcsrv.watson.ibm.com/Tmsa.html`.

J. Sequence formats

One of the simplest examples of multiple sequence input is sequence data in the **Pearson/Fasta format**. In this format, the first line begins with a greater-than character ('>'). This character indicates the start of a sequence. The first word after the the '>' is the name of the sequence; any other words on this line are considered to be a comment/title for the sequence. All lines until the next line beginning with a '>' (or until there are no more lines) contain the actual sequence data (no spaces or numbers) of the first sequence.

Here is one example of sequence data in the **Pearson/Fasta format**:

```
> locus1
GCGCGCGCGCGCGCGCGCGCGCGCGCGCGCGCGCGCGCGCGCGCGCGCGCGC
GCGCGCGCGCGCGCGCGCGCGCGCGCGCGCGCGCGCGCGCGCGCGCGCGCGC
GCGCGCGCGC

> locus2
ATATATATATATATATATATATATATATATATATATATATATATATATAT
ATATATATATATATATATATATATATATATATATATATATATATATATAT
ATATATATAT

> locus3
ACGTACGTACGT
```

Lines should be no longer than 80 characters. The next line beginning with a '>' (if there are any more lines) marks the start of the second sequence. Again, the first word after the '>' is the name of the second sequence. Any additional words in that line are the a comment/title for the second sequence. This second line beginning with a '>' should be followed by lines containing the sequence data until another line beginning with a '>', (or until there are no more lines). Additional sequences can follow the second sequence in a similar fashion.

K. VMD

K.1. Installation

VMD has been available at http://www.ks.uiuc.edu/Research/vmd. In the past, versions for IRIX, Linux, MacOS X, Solaris, and Windows were available, along with documentation for installation and use of VMD.

Some installation hints:

- don't download the CUDA version unless you are sure you have CUDA

- if installing on a Mac, familiarize yourself with how to install an application which is distributed as a disk images (a .dmg file)[1]

K.2. Using VMD

K.2.1. Viewing a molecule

To view a molecule, click on "File" to access the File pull-down menu. Select "New Molecule" to access the molecule file browser. Click the "Browse" button to bring up a file browser which you can use to select the file of interest. Press the "Load" button to visualize the data in the file.

---

[1]Drag and drop the application from the virtual drive into your Applications folder. Then, open your Applications folder, and double-click on the application icon in the Applications folder.

K.2.2. Using the mouse to adjust the view of the molecule(s)

Use the keyboard to select the mode in which the mouse will operate as follows:

- press the **r** key to rotate the molecule

- press the **t** key to translate the molecule

- press the **s** key to scale the molecule

See the 'Hot keys' section of the VMD user guide for more details.

With each of these modes, you are moving the scene (i.e., the view) and the molecule (if the molecule is not fixed (selected in the VMD Main window under molecule)) or the scene only (if the molecule is fixed). To move the molecule (the atomic coordinates), use Mouse; Move; Molecule.

Press **CTRL-R** to reset the view of a molecule.

With two molecules, there are two approaches to super-imposing them (after first loading each molecule individually):

toggle the fixed setting on one molecule and then translate the other
move the atomic coordinates of one using Mouse; Move; Molecule

K.2.3. Modifying the representation of a molecule

These commands are accessed by opening the Graphical Representations window (from the VMD Main window, use the Graphics pull-down to select Representations).

Use Display - Axes - buttons to remove or relocate the xyz axis indicator from the view.

Use the "Coloring method" option on the Graphical Representations window to select different coloring schemes (use in conjunction with Graphics, Colors). Coloring schemes can be based on a number of properties, including:

beta:

> color based on the beta (anisotropic temperature) value of the PDB file

chain:

> color each atom based on which chain it belongs to

colorID:

> use a user-specified color index (from 0 to 15)

name:

> each atom receives a different color based on its identity in the periodic table. You can determine which atoms correspond to which color by using the Color Form (go to the "VMD main" window, select the Graphics pull-down menu, and select Colors. The colors by default include:
>
> H - white
> O - red
> N - blue
> C - cyan (for the non-artistic, this is a light blue-green)
> S - yellow
> P - tan

index:

> color based on atom index in PDB file

resname:

the color of each atom is based on the name of the amino acid residue to which it belongs

structure:

the color of each atom is based on whether it is a component of alpha helix, beta sheet, or random coil

type:

uses the 'Type' category

Use the "Drawing method" option to select style of drawing for the given representation:

H bonds
ribbons
tube
cartoons: show nice depiction including secondary structure cartoons
VDW: van der Walls radii
CPK: a nice ball and stick representation
etc.

selections

Use the "Selections" tab to select specific components of the molecule to view.

Specification of a residue is used to identify a set of atoms (typically, a residue is defined as an individual amino acid or nucleotide residue in the context of a polymer – e.g., a polypeptide or nucleic acid). Note that selections are case-sensitive ("chain a" is not the same as "chain A"). Note also that you can see the macro definition for many of the 'singleword' macros under the 'Selections' tab of the 'Graphical Representations' window.

Every atom in a model has a specific index number. This number is unique to that atom within the context of the

VMD selection syntax	atoms selected
protein	only protein
backbone	only backbone of protein
resname HEM	only the heme residues
chain A or chain B	chain A and chain B
residue 23 to 74	residues 23 to 74
residue 0 5	residue 0 and residue 5
resname ALA PHE ASP	all atoms in ALA, PHE, or ASP
acidic	residues ASP and GLU
basic	residues ARG HIS LYS
buried	residues ALA LEU VAL ILE PHE CYS MET TRP
charged	same as basic or acidic
hydrophobic	ALA LEU VAL ILE PRO PHE MET TRP
polar	same as 'protein and not hydrophobic'

Table K.1. Examples of selections

model under consideration. Note that residue numbers and index numbers are *not* the same numbers.

More complex selections can be constructed with binary operators; e.g., not protein and not resname HEM would correspond to all items that are not part of the covalent structure of the protein and that are not heme groups.

K.2.4. Representations

More than one view of a given molecule or a portion thereof can be shown at the same time. To create a new representation of the molecule, click the "Create rep" button in the "Graphical representation" window. The new representation can have a totally different color scheme, different drawing scheme, different atoms selected, etc.

K.2.5. Identifying residues or regions of a molecule

Use the keyboard to select the mouse mode by pressing "0" (the "zero" key). In this mode, clicking on an atom will cause the details associated with the atom (e.g., the chain to which it belongs, the amino acid residue number in the chain, etc.) to be displayed in the "VMD console" window (on some operating systems, the window may not actually be titled the "VMD console" window; for example, on Windows machines, you may want to look for an "MS-DOS prompt" window or a window with a "C:\" icon)..

Note that the **name** of an atom can be determined by color if the coloring scheme is set appropriately.

K.2.6. Determining distances

Use the keyboard to select Label → Bond mode by pressing "2". In this mode, clicking on two atoms sequentially toggles on/off a bond distance label (default units are Angstroms).

The display can get cluttered up pretty quickly after several such distances have been measured. Use the '=' button to reset the view. To remove labels, go to the 'Main' window, select the Graphics pulldown, select Labels, and then delete each label. Unfortunately, the dashed lines representing the intraatomic axis and the corresponding length annotations are still present...

K.2.7. Saving an image

1. From the VMD Main window, select File; Render

2. Choose the "snapshot" method

3. On Unix systems, this will result in the file being saved as an RGB file or a TGA file; on Windows systems, this will result in the file being saved as a BMP file

note(s):

- the snapshot renderer saves *exactly* what is already showing in the display window (e.g., if another window overlaps the display window, the contents of that window will be rendered)

K.2.8. Viewing images

For BMP images, try using newer versions of mozilla. RGB files can be converted to other formats using a utility such as Imagemagick.

You may want to change the background color. Do this by selecting Graphics → Colors. Then one can either select 'Background' or click on the "Display" category and then select the background name.

K.2.9. Printing an image

When you click on the "File" pull-down menu on the VMD main window, "Print" is *not* listed as an option. This is not necessarily cause to panic. Consider saving the image (see "Saving an image", above) and then opening the image in an image browsing program (e.g., firefox or gimp) and then printing the image.

K.3. Plugins

VMD functionality can be increased with a number of plugins/extensions (see http://www.ks.uiuc.edu/Research/

`vmd/plugins/`). Some that may be of interest include:

Contact Map
Dowser

 - add water to a structure

Membrane
Molefacture

- an interface for editing molecules

- automatic hydrogen placement

MultiSeq
RamaPlot
Salt Bridges
Solvate

 - see also Dowser

K.4. Odds and ends

Where are the hydrogens in the pdb file?

- in many cases, the resolution of crystal structures isn't sufficient to identify the location of hydrogen atoms

- some software is designed to guess where H's might be (e.g., reduce (`http://kinemage.biochem.duke.edu/software/reduce.php`), openbabel, vega, whatif, pdb2pqr (`http://pdb2pqr.sourceforge.net/`), vmd molefacture plugin)

L. Reading a scientific paper

This content is from "How to Read a Scientific Paper" (John W. Little and Roy Parker; University of Arizona) (http://www.biochem.arizona.edu/classes/bioc568/papers.htm) and is reproduced with permission from the authors.[1]

L.1. How to Read a Scientific Paper

The main purpose of a scientific paper is to report new results, usually experimental, and to relate these results to previous knowledge in the field. Papers are one of the most important ways that we communicate with one another.

In understanding how to read a paper, we need to start at the beginning with a few preliminaries. We then address the main questions that will enable you to understand and evaluate the paper.

1. How are papers organized?

2. How do I prepare to read a paper, particularly in an area not so familiar to me?

3. What difficulties can I expect?

---

[1]The web page content, accessed on 2009-10-23, was reformatted for print (images present in the original web page are omitted here and hyperlink text content is not underlined and is replaced with the actual URL where appropriate). Otherwise, the text is the same text as in the original web page.

4. How do I understand and evaluate the contents of the paper?

L.1.1. 1. Organization of a paper

In most scientific journals, scientific papers follow a standard format. They are divided into several sections, and each section serves a specific purpose in the paper. We first describe the standard format, then some variations on that format.

A paper begins with a short **Summary** or **Abstract**. Generally, it gives a brief background to the topic; describes concisely the major findings of the paper; and relates these findings to the field of study. As will be seen, this logical order is also that of the paper as a whole.

The next section of the paper is the **Introduction**. In many journals this section is not given a title. As its name implies, this section presents the background knowledge necessary for the reader to understand why the findings of the paper are an advance on the knowledge in the field. Typically, the Introduction describes first the accepted state of knowledge in a specialized field; then it focuses more specifically on a particular aspect, usually describing a finding or set of findings that led directly to the work described in the paper. If the authors are testing a hypothesis, the source of that hypothesis is spelled out, findings are given with which it is consistent, and one or more predictions are given. In many papers, one or several major conclusions of the paper are presented at the end of this section, so that the reader knows the major answers to the questions just posed. Papers more descriptive or comparative in nature may begin with an introduction to an area which interests the authors, or the need for a broader database.

The next section of most papers is the **Materials and Methods**. In some journals this section is the last one. Its purpose is to describe the materials used in the experiments and the methods by which the experiments were carried out. In principle, this description should be detailed enough to allow other researchers to replicate the work. In practice, these descriptions are often highly compressed, and they often refer back to previous papers by the authors.

The third section is usually **Results**. This section describes the experiments and the reasons they were done. Generally, the logic of the Results section follows directly from that of the Introduction. That is, the Introduction poses the questions addressed in the early part of Results. Beyond this point, the organization of Results differs from one paper to another. In some papers, the results are presented without extensive discussion, which is reserved for the following section. This is appropriate when the data in the early parts do not need to be interpreted extensively to understand why the later experiments were done. In other papers, results are given, and then they are interpreted, perhaps taken together with other findings not in the paper, so as to give the logical basis for later experiments.

The fourth section is the **Discussion**. This section serves several purposes. First, the data in the paper are interpreted; that is, they are analyzed to show what the authors believe the data show. Any limitations to the interpretations should be acknowledged, and fact should clearly be separated from speculation. Second, the findings of the paper are related to other findings in the field. This serves to show how the findings contribute to knowledge, or correct the errors of previous work. As stated, some of these logical arguments are often found in the Results when it is necessary to clarify why later experiments were carried out. Although you might argue that in this case the discussion material should be presented in the Introduction, more often you cannot grasp its significance until the first part of Results is given.

Finally, papers usually have a short **Acknowledgements**

section, in which various contributions of other workers are recognized, followed by a **Reference** list giving references to papers and other works cited in the text.

Papers also contain several **Figures and Tables**. These contain data described in the paper. The figures and tables also have legends, whose purpose is to give details of the particular experiment or experiments shown there. Typically, if a procedure is used only once in a paper, these details are described in Materials and Methods, and the Figure or Table legend refers back to that description. If a procedure is used repeatedly, however, a general description is given in Materials and Methods, and the details for a particular experiment are given in the Table or Figure legend.

Variations on the organization of a paper

In most scientific journals, the above format is followed. Occasionally, the Results and Discussion are combined, in cases in which the data need extensive discussion to allow the reader to follow the train of logic developed in the course of the research. As stated, in some journals, Materials and Methods follows the Discussion. In certain older papers, the Summary was given at the end of the paper.

The formats for two widely-read journals, *Science* and *Nature*, differ markedly from the above outline. These journals reach a wide audience, and many authors wish to publish in them; accordingly, the space limitations on the papers are severe, and the prose is usually highly compressed. In both journals, there are no discrete sections, except for a short abstract and a reference list. In *Science*, the abstract is self-contained; in *Nature*, the abstract also serves as a brief introduction to the paper. Experimental details are usually given either in endnotes (for *Science*) or Figure and Table legends and a short Methods section (in *Nature*). Authors often try to circumvent length limitations by putting as much material as possible in these places. In addition, an increasingly common practice is to put a substantial fraction of the less-important material, and much of the methodology, into Supplemental Data that can be accessed online.

Many other journals also have length limitations, which similarly lead to a need for conciseness. For example, the *Proceedings of the National Academy of Sciences (PNAS)* has a six-page limit; *Cell* severely edits many papers to shorten them, and has a short word limit in the abstract; and so on.

In response to the pressure to edit and make the paper concise, many authors choose to condense or, more typically, omit the logical connections that would make the flow of the paper easy. In addition, much of the background that would make the paper accessible to a wider audience is condensed or omitted, so that the less-informed reader has to consult a review article or previous papers to make sense of what the issues are and why they are important. Finally, again, authors often circumvent page limitations by putting crucial details into the Figure and Table legends, especially when (as in *PNAS*) these are set in smaller type. Fortunately, the recent widespread practice of putting less-critical material into online supplemental material has lessened the pressure to compress content so drastically, but it is still a problem for older papers.

L.1.2. 2. Reading a scientific paper

Although it is tempting to read the paper straight through as you would do with most text, it is more efficient to organize the way you read. Generally, you first read the Abstract in order to understand the major points of the work. The extent of background assumed by different authors, and allowed by the journal, also varies as just discussed.

One extremely useful habit in reading a paper is to read the Title and the Abstract and, before going on, review in your mind what you know about the topic. This serves several purposes. First, it clarifies whether you in fact know enough background to appreciate the paper. If not, you might choose to read the background in a review or textbook, as appropriate.

Second, it refreshes your memory about the topic. Third, and perhaps most importantly, it helps you as the reader integrate the new information into your previous knowledge about the topic. That is, it is used as a part of the self-education process that any professional must continue throughout his/her career.

If you are very familiar with the field, the Introduction can be skimmed or even skipped. As stated above, the logical flow of most papers goes straight from the Introduction to Results; accordingly, the paper should be read in that way as well, skipping Materials and Methods and referring back to this section as needed to clarify what was actually done. A reader familiar with the field who is interested in a particular point given in the Abstract often skips directly to the relevant section of the Results, and from there to the Discussion for interpretation of the findings. This is only easy to do if the paper is organized properly.

Codewords

Many papers contain shorthand phrases that we might term 'codewords', since they have connotations that are generally not explicit. In many papers, not all the experimental data are shown, but referred to by "(data not shown)". This is often for reasons of space; the practice is accepted when the authors have documented their competence to do the experiments properly (usually in previous papers). Two other codewords are "unpublished data" and "preliminary data". The former can either mean that the data are not of publishable quality or that the work is part of a larger story that will one day be published. The latter means different things to different people, but one connotation is that the experiment was done only once.

L.1.3. 3. Difficulties in reading a paper

Several difficulties confront the reader, particularly one who is not familiar with the field. As discussed above, it may be necessary to bring yourself up to speed before beginning a paper, no matter how well written it is. Be aware, however, that although some problems may lie in the reader, many are the fault of the writer.

One major problem is that many papers are poorly written. Some scientists are poor writers. Many others do not enjoy writing, and do not take the time or effort to ensure that the prose is clear and logical. Also, the author is typically so familiar with the material that it is difficult to step back and see it from the point of view of a reader not familiar with the topic and for whom the paper is just another of a large stack of papers that need to be read.

Bad writing has several consequences for the reader. First, the logical connections are often left out. Instead of saying why an experiment was done, or what ideas were being tested, the experiment is simply described. Second, papers are often cluttered with a great deal of jargon. Third, the authors often do not provide a clear road-map through the paper; side issues and fine points are given equal air time with the main logical thread, and the reader loses this thread. In better writing, these side issues are relegated to Figure legends, Materials and Methods, or online Supplemental Material, or else clearly identified as side issues, so as not to distract the reader.

Another major difficulty arises when the reader seeks to understand just what the experiment was. All too often, authors refer back to previous papers; these refer in turn to previous papers in a long chain. Often that chain ends in a paper that describes several methods, and it is unclear which was used. Or the chain ends in a journal with severe space limitations, and the description is so compressed as to be unclear. More often, the descriptions are simply not well-written, so that it is ambiguous what was done.

Other difficulties arise when the authors are uncritical about their experiments; if they firmly believe a particular model, they may not be open-minded about other possibilities. These may not be tested experimentally, and may even go unmentioned in the Discussion. Still another, related problem is that many authors do not clearly distinguish between fact and speculation, especially in the Discussion. This makes it difficult for the reader to know how well-established are the "facts" under discussion.

One final problem arises from the sociology of science. Many authors are ambitious and wish to publish in trendy journals. As a consequence, they overstate the importance of their findings, or put a speculation into the title in a way that makes it sound like a well-established finding. Another example of this approach is the "Assertive Sentence Title", which presents a major conclusion of the paper as a declarative sentence (such as "LexA is a repressor of the recA and lexA genes"). This trend is becoming prevalent; look at recent issues of Cell for examples. It's not so bad when the assertive sentence is well-documented (as it was in the example given), but all too often the assertive sentence is nothing more than a speculation, and the hasty reader may well conclude that the issue is settled when it isn't.

These last factors represent the public relations side of a competitive field. This behavior is understandable, if not praiseworthy. But when the authors mislead the reader as to what is firmly established and what is speculation, it is hard, especially for the novice, to know what is settled and what is not. A careful evaluation is necessary, as we now discuss.

Type of research	Question asked:
Descriptive	What is there? What do we see?
Comparative	How does it compare to other organisms? Are our findings general?
Analytical	How does it work? What is the mechanism?

L.1.4. 4. Evaluating a paper

A thorough understanding and evaluation of a paper involves answering several questions:

a. What questions does the paper address?

b. What are the main conclusions of the paper?

c. What evidence supports those conclusions?

d. Do the data actually support the conclusions?

e. What is the quality of the evidence?

f. Why are the conclusions important?

a. What questions does the paper address?

Before addressing this question, we need to be aware that research in biochemistry and molecular biology can be of several different types:

Descriptive research often takes place in the early stages of our understanding of a system. We can't formulate hypotheses about how a system works, or what its interconnections are, until we know what is there. Typical descriptive approaches in molecular biology are DNA sequencing and DNA microarray approaches. In biochemistry, one could regard x-ray crystallography as a descriptive endeavor.

179

Comparative research often takes place when we are asking how general a finding is. Is it specific to my particular organism, or is it broadly applicable? A typical comparative approach would be comparing the sequence of a gene from one organism with that from the other organisms in which that gene is found. One example of this is the observation that the actin genes from humans and budding yeast are 89% identical and 96% similar.

Analytical research generally takes place when we know enough to begin formulating hypotheses about how a system works, about how the parts are interconnected, and what the causal connections are. A typical analytical approach would be to devise two (or more) alternative hypotheses about how a system operates. These hypotheses would all be consistent with current knowledge about the system. Ideally, the approach would devise a set of experiments todistinguish among these hypotheses. A classic example is the Meselson-Stahl experiment.

Of course, many papers are a combination of these approaches. For instance, researchers might sequence a gene from their model organism; compare its sequence to homologous genes from other organisms; use this comparison to devise a hypothesis for the function of the gene product; and test this hypothesis by making a site-directed change in the gene and asking how that affects the phenotype of the organism and/or the biochemical function of the gene product.

Being aware that not all papers have the same approach can orient you towards recognizing the major questions that a paper addresses.

What are these questions? In a well-written paper, as described above, the Introduction generally goes from the general to the specific, eventually framing a question or set of questions. This is a good starting place. In addition, the results of experiments usually raise additional questions, which the authors may attempt to answer. These questions usually become evident only in the Results section.

b. What are the main conclusions of the paper?

This question can often be answered in a preliminary way by studying the abstract of the paper. Here the authors high-light what they think are the key points. This is not enough, because abstracts often have severe space constraints, but it can serve as a starting point. Still, you need to read the paper with this question in mind.

c. What evidence supports those conclusions?

Generally, you can get a pretty good idea about this from the Results section. The description of the findings points to the relevant tables and figures. This is easiest when there is one primary experiment to support a point. However, it is often the case that several different experiments or approaches combine to support a particular conclusion. For example, the first experiment might have several possible interpretations, and the later ones are designed to distinguish among these.

In the ideal case, the Discussion begins with a section of the form "Three lines of evidence provide support for the conclusion that... First, ...Second,... etc." However, difficulties can arise when the paper is poorly written (see above). The authors often do not present a concise summary of this type, leaving you to make it yourself. A skeptic might argue that in such cases the logical structure of the argument is weak and is omitted on purpose! In any case, you need to be sure that you understand the relationship between the data and the conclusions.

d. Do the data actually support the conclusions?

One major advantage of doing this is that it helps you to evaluate whether the conclusion is sound. If we assume for the moment that the data are believable (see next section), it still might be the case that the data do not actually support the conclusion the authors wish to reach. There are at least two different ways this can happen:

1. The logical connection between the data and the interpretation is not sound

2. There might be other interpretations that might be consistent with the data.

One important aspect to look for is whether the authors take multiple approaches to answering a question. Do they have multiple lines of evidence, from different directions, supporting their conclusions? If there is only one line of evidence, it is more likely that it could be interpreted in a different way; multiple approaches make the argument more persuasive.

Another thing to look for is implicit or hidden assumptions used by the authors in interpreting their data. This can be hard to do, unless you understand the field thoroughly.

e. What is the quality of that evidence?

This is the hardest question to answer, for novices and experts alike. At the same time, it is one of the most important skills to learn as a young scientist. It involves a major reorientation from being a relatively passive consumer of information and ideas to an active producer and critical evaluator of them. This is not easy and takes years to master. Beginning scientists often wonder, "Who am I to question these authorities? After all the paper was published in a top journal, so the authors must have a high standing, and the work must have received a critical review by experts." Unfortunately, that's not always the case. In any case, developing your ability to evaluate evidence is one of the hardest and most important aspects of learning to be a critical scientist and reader.

How can you evaluate the evidence?

First, you need to understand thoroughly the methods used in the experiments. Often these are described poorly or not at all (see above). The details are often missing, but more importantly the authors usually assume that the reader has a general knowledge of common methods in the field (such as immunoblotting, cloning, genetic methods, or DNase I footprinting). If you lack this knowledge, as discussed above you have to make the extra effort to inform yourself about the basic methodology before you can evaluate the data.

Sometimes you have to trace back the details of the methods if they are important. The increasing availability of journals on the Web has made this easier by obviating the need to find a hard-copy issue, e.g. in the library. A comprehensive listing of journals relevant to this course, developed by the Science Library, allows access to most of the listed volumes from any computer at the University; a second list at the Arizona Health Sciences Library includes some other journals, again from University computers.

Second, you need to know the **limitations** of the methodology. Every method has limitations, and if the experiments are not done correctly they can't be interpreted.

For instance, an immunoblot is not a very quantitative method. Moreover, in a certain range of protein the signal increases (that is, the signal is at least roughly "linear"), but above a certain amount of protein the signal no longer increases. Therefore, to use this method correctly one needs a standard curve that shows that the experimental lanes are in a linear range. Often, the authors will not show this standard curve, but they should state that such curves were done.

If you don't see such an assertion, it could of course result from bad writing, but it might also not have been done. If it wasn't done, a dark band might mean "there is this much protein or an indefinite amount more".

Third, importantly, you need to distinguish between what the data show and what the authors **say** they show. The latter is really an interpretation on the authors' part, though it is generally not stated to be an interpretation. Papers usually state something like "the data in Fig. x show that ...". This is the authors' interpretation of the data. Do you interpret it the same way? You need to look carefully at the data to ensure that they really do show what the authors say they do. You can only do this effectively if you understand the methods and their limitations.

Fourth, it is often helpful to look at the original journal, or its electronic counterpart, instead of a photocopy. Particularly for half-tone figures such as photos of gels or autoradiograms, the contrast is distorted, usually increased, by photocopying, so that the data are misrepresented.

Fifth, you should ask if the proper controls are present. Controls tell us that nature is behaving the way we expect it to under the conditions of the experiment (see The Importance of Control Experiments (http://www.biochem.arizona.edu/classes/bioc568/Controls.htm) for more details). If the controls are missing, it is harder to be confident that the results really show what is happening in the experiment. You should try to develop the habit of asking "where are the controls?" and looking for them.

f. Why are the conclusions important?

Do the conclusions make a significant advance in our knowledge? Do they lead to new insights, or even new research directions?

Again, answering these questions requires that you understand the field relatively well.

M. Protein assays

M.1. Protein concentration assays

Since proteins are the workhorses of the cell, scientists are frequently interested in the relative abundance, activity, or biochemical characteristics of a protein in a sample. Samples may consist of relatively purified protein from a protein purification protocol or may consist of crude cell or tissue homogenates ("lysates"). In all cases, it is important to determine the amount and concentration of protein in the sample before subjecting it to further manipulations. There are several means of estimating protein concentration in a sample, each with its pros and cons; several of the more commonly utilized techniques are briefly described below.

Spectrophotometry
Absorbance assays depending on a protein's intrinsic UV absorbance (typically at 280 nm and 205 nm) are occasionally used to evaluate protein concentration. These assays are excellent for use with relatively pure protein solutions; however, the solution must be free of other UV-absorbing substances (e.g., absorbance at 205 and 280 nm is subject to interference from nucleic acid or lipid components in a sample).

Bradford assay
The Bradford total protein quantitation assay [5], and several commercial modifications thereof, are colorimetric assays based on the tendency of the dye Coomassie G-250 to shift absorbance from 465 nm to 595 nm (there is a simultaneous color change of the reagent from red/brown to blue)

when the reagent binds proteins in an acidic solution. The mechanism of the reaction is based on an anionic form of the dye; this form interacts primarily with arginine residues and, more weakly, with histidine, lysine, tyrosine, tryptophan, and phenylalanine residues. The reaction reaches a relatively stable endpoint so valid absorption measurements can be made over the course of the hour following equilibration of the dye and protein.

The Bradford assay is subject to interference from some compounds (see the Bradford protocol).

In any protein assay the best protein to use as a standard is a purified preparation of the protein being assayed. In the absence of such an absolute reference protein another protein must be selected as a relative standard. The best relative standard to use is one which gives a color yield similar to that of the protein being assayed. Selecting such a protein standard is generally done empirically. Alternatively, if only relative protein values are desired, any purified protein may be selected as a standard. If a direct comparison of two different protein assays is being performed, the same standard should be used for both procedures. With the Bradford protein assay, the dye color development is significantly greater with albumin than with most other proteins, including gamma-globulin. Therefore, although an albumin standard is commonly used, for a color response that is typical of many proteins, the gamma-globulin standard is appropriate. (adapted from BioRad literature)

BCA assay
Both the bicinchoninic acid (BCA) assay, as well as a more time-consuming assay termed the "Lowry assay", are colorimetric assays based on the reduction of Cu^{2+} to Cu^{1+} by amides. BCA is a highly selective and sensitive detection agent for Cu^{1+}. The macromolecular structure of the protein, the peptide bond number, and the presence of cysteine, cystine, tryptophan, and tyrosine contribute to the development of color in the BCA assay.

The BCA assay is subject to interference from some compounds, particularly reducing agents and copper chelators.

Other assays

Miscellaneous other assays are also used in assessing protein concentration. The Biuret assay is approximately 100-fold less sensitive than the Bradford assay (a more sensitive Biuret assay protocol was reported by Matsushita et al. [18]). Pyrogallol red-molybdate is infrequently used, with a reported sensitivity of 5-50 µg/10 µL. The "Nanoorange" reagent is a recently introduced, highly sensitive protein detection reagent.

The standard curve

A standard curve describes the response of an assay to introduction of varying amounts of a standard, a compound representative of the sample to be assayed. The resulting data can be represented as a two-dimensional X-Y plot with the X axis representing the response and the Y axis representing the value of the standard. For many assays, the relationship between the response of the assay and the amount of standard is linear over a particular range. This is frequently considered the useful, or "working", range of the assay. A linear regression can be conducted to mathematically define the linear portion of the curve. Clearly, the more data points (in the linear range) included in the regression, the higher the confidence the investigator has in the final regression formula. With more sophisticated computing resources, all data points can be included and the appropriate non-linear regression (if known) used. Regardless, the resulting formula, $y=f(x)$, can be used to evaluate sample data. Introducing the assessed response value of a sample, x_{sample}, into the formula, yields the corresponding y value, typically representing the extrapolated concentration or mass of the sample.

Although a linear regression can be calculated by hand by the statistically competent, many calculators have statistical functions including linear regression which make the process much more rapid. It may be worth checking for these functions on your calculator or in your calculator manual. Computer programs are freely available (e.g., gnuplot) which have the capability to do relatively sophisticated regression analyses. Alternatively, a variety of online sites provide tools to conduct linear regression.

N. The Bradford protein assay

N.1. Introduction

The Bradford total protein quantitation assay [5], and several commercial modifications thereof, are colorimetric assays based on the tendency of the dye Coomassie G-250 to shift absorbance from 465 nm to 595 nm (there is a simultaneous color change of the reagent from red/brown to blue) when the reagent binds proteins in an acidic solution. Although it is common to simply use absorption at 450 as the endpoint of the assay, the sensitivity and linearity of the assay can be improved by evaluating the ratio of absorption at 595 to absorption at 465 for each sample [29]. The mechanism of the reaction is based on an anionic form of the dye; this form interacts primarily with arginine residues and, more weakly, with histidine, lysine, tyrosine, tryptophan, and phenylalanine residues. The reaction reaches a relatively stable endpoint so valid absorption measurements can be made over the course of the hour following equilibration of the dye and protein.

The Bradford assay does not have a wide linear range. Thus, it may be necessary to prepare sample dilutions prior to analysis.

The Bradford assay is subject to interference from some compounds. The following compounds are not likely to interfere at the indicated concentrations:[1,2,3]

β-mercaptoethanol, 1M
dithiothreitol, 5 mM - 1M
EDTA, 0.1M
EGTA, 0.002 - 0.05M
ethanol, 10%
glucose, 20%, 1M
HEPES, 0.1M
SDS, 0.1%
Triton X-100, 0.1 - 0.125%
urea, 3 - 6M

In any protein assay the ideal standard is a purified preparation of the protein or protein mix being assayed. Since color yield varies with protein with the Bradford assay, the choice of this standard is significant if precise quantitation is desired. However, such a reference standard often isn't available. In such a situation, another protein (or set of proteins) is used as a relative standard. The ideal relative standard is that protein or protein mix which gives a color yield similar to that of the protein being assayed. Selecting such a protein standard is generally done empirically. Alternatively, if only relative protein values are desired, any purified protein may be selected as a standard.

N.2. Procedure

If the samples being evaluated will be needed at a later time, they should be kept at 4°C. Ensure an ice bucket or a chilled tube holder is on hand before beginning.

[1] Bio-Rad, Bio-Rad Protein Assay Rev C
[2] Thermo Scientific, 2009, Thermo Scientific Pierce Protein Assay Technical Handbook
[3] DA Thompson, unpublished data

1. If BSA stock solutions are not provided, prepare BSA (bovine serum albumin) stocks as described below:

   3 mg/mL BSA

   > example: add 30 mg BSA to 10 mL water

   at least 20 µL of 1 mg/mL BSA

   > example: add 16.7 µL of 3 mg/mL BSA solution to 33.3 µL water

   at least 30 µL of 0.1 mg/mL BSA

   > example: add 3 µL of 1 mg/mL BSA solution to 27 µL water

2. Prepare (label) a set of tubes according to Figure N.1

3. Add water to *each* tube following Figure N.1[4]

4. Add protein (sample or standard) to *each* tube according to Figure N.1

5. Add 200 µL dye ('Bradford reagent') to each tube; vortex vigorously; decant into plastic 1 mL cuvets

6. Wait at least 10 min (but not more than 1 h) at room temperature after adding dye

7. Zero the spectrophotometer (see Appendix P) with water at A_{590}

   > - after zeroing at A_{450}, record the baseline A_{450} value (the value with water) so that subsequent

---

[4]a meditation on the word "each": here, 'each' is used as an adjective. Merriam-Webster describes the etymology of each as deriving originally "from West Germanic *aiw (ever, always)" and defines 'each' as "being one of two or more distinct individuals having a similar relation and often constituting an aggregate" (http://www.merriam-webster.com/dictionary/each ; 2008-03-12). In the sense 'each' is used here (i.e., in this Bradford assay protocol), the idea is that you should have completed the step for **all** of the tubes/samples/standards **before** proceeding to the next step. Have you?

A_{450} measurements can be blank-corrected

8. Measure A_{590} and A_{450} for each sample:

   - directly before measuring, vortex or invert two or three times to mix the sample

   - decant the sample from the microcentrifuge tube into a cuvet

   - record each absorbance value

N.2.1. calculations

1. Calculate A_{590}/A_{450} for each sample

2. Construct a standard curve

3. Use linear regression to mathematically define the observed relationship between protein concentration and A_{590}/A_{450}

4. Evaluate the concentration of the unknown sample(s)

Standard (μg/mL)	μL standard	μL water	μL dye	A_{590}/A_{450} (1)	A_{590}/A_{450} (2)
0		800	200		
0.2	2 (0.1 mg/mL)	798	200		
1	10 (0.1 mg/mL)	790	200		
5	5 (1 mg/mL)	795	200		
10	10 (1 mg/mL)	790	200		
60	20 (3 mg/mL)	780	200		
Sample					
A	0.1*	799.9	200		
B	2	798	200		
C	8	792	200		

* Is it practical to measure 0.1 microliter? If not, and if this is the desired volume of sample, how can the situation be resolved?

Figure N.1. Samples and standards

O. Densitometry

Several software packages are available for analyzing images such as those of gels. These include tnimage (`http://brneurosci.org/tnimage.html`), a Linux densitometry package, NIH Image (`http://rsb.info.nih.gov/nih-image/`), an image processing program for the Macintosh, Image/J (`http://rsb.info.nih.gov/ij/`), a Java-based program somewhat similar to NIH Image, and Scion Image (`http://www.scioncorp.com/`), a Windows version of NIH Image.

O.1. Using ImageJ

Start ImageJ (in Linux, 'java -jar ij.jar')

Open the image: File → Open to import a TIFF or JPEG image

Use the "Magnifying glass" tool to zoom in (right-click (or alt-click) to zoom out)

Use Image → Rotate to rotate the image so that the lanes run vertically

If desired, use Image → Crop to crop the image

To measure distances (e.g., to calculate Rf values), use the "Line tool" to draw a line between the points of interest and use the Analyze → Measure tool to determine the length between the two points

For densitometry measurements,

- use a selection tool (e.g., the rectangle tool) to select a region of interest

- determine the mean grey value: Analyze → Measure (if the mean value isn't given use Analyze → Set measurements to specify which measurements are recorded)

- remember to determine the background mean grey value – subtract it from other values to obtain the values attributable to the protein band(s)

O.2. Using Scion Image

1. Open the image: click the File pull-down and select Import to import a TIFF image

2. Use the "Magnifying glass" tool to adjust the view

3. Calibrate using loading standards and analyze sample bands of interest.

   a. Analyze → Reset

   b. Use a selection tool to record the mean gray value of each standard. Starting with the lowest standard, select the band. Click Analyze and select Measure (Ctrl - 1). Repeat with remaining standards, progressing to the highest standard.

   c. Click Analyze and select Calibrate (use a straight-line fit)

   d. Use the selection tool to record the mean gray value of each sample band of interest. Click Analyze and select Calibrate to obtain the mean gray value.

P. Spectrophotometer use

P.1. Genesys 20 Vis spectrophotometer

The wavelength range of the Genesys 20 Vis spectrophotometer is 325 nm to 1100 nm.

1. Turn the instrument on (the switch is located on the rear LHS). The machine should perform power-on sequence (note: ensure that cell holder is empty during the power-on sequence).

2. For a full warm-up, allow at least 30 min before use.

3. Perform Abs/Trans measurements by following the steps below:

   - press [A/T/C] to select Abs or Trans mode

   - press [nm up/down] to select wavelength

   - insert blank into cell holder and close the sample door[1]

   - press [0 ABS/100%T] to set the blank zero

   - remove blank; insert the cuvet containing the sample and close the sample door

   - measurement appears on the display

While the specifics of spectrophotometer use vary from instrument to instrument (e.g., the steps used with the Gensys

---

[1]Ensure you position the cell so light passes through the clear walls of the cell.

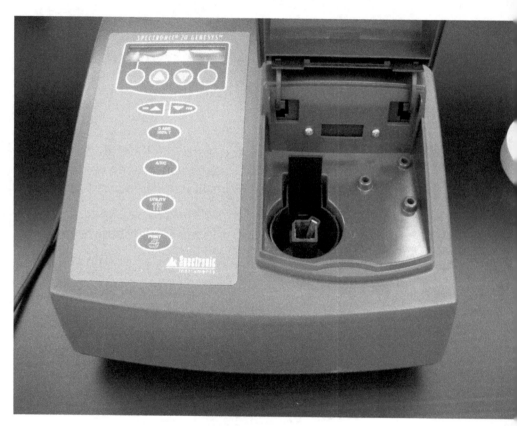

Figure P.1. The sample compartment and control panel of a spectrophotometer

Figure P.2. A cuvet

20 Vis are below), the basics are pretty similar from instrument to instrument:

- turn the instrument on, allowing time to complete any on-board diagnostics and time for the lamp(s) to "warm up"

- select the desired wavelength

- zero the instrument on a "blank":

  • fill a cuvet (Figure P.2) with the solvent in which the sample is dissolved (fill to a position at least 0.5 cm above the position corresponding to the light path)

  • place the cuvet in the spectrophotometer cuvet holder, ensuring that the transparent sides of the cuvet are situated so that they are perpendicular to the light path (i.e., so that the light path passes through the transparent faces) (note the arrow in Figure P.3)

  • press the 'zero' or 'clear' button on the instrument (if no such control exists, record the absorbance value – this is the value which will be subtracted from all other

201

Figure P.3. The sample compartment of a spectrophotometer

measurements)

- measure the absorbance of the sample, preparing and placing the cuvet in the same manner as for the "blank"

P.2. The Genesys 20 Vis spectrophotometer

note(s): the wavelength range of the Gensys 20 Vis is 325 - 1100 nm

202

1. Turn the instrument on (the switch is located on the rear left-hand side). The machine should perform a power-on sequence.

   > note: ensure that cell holder is empty during power-on sequence

2. For full warm-up, allow at least 30 min before use

3. Abs/Trans measurements:

   - press [A/T/C] to select Abs or Trans mode
   - press [nm up/down] to select wavelength
   - insert blank into cell holder; close sample door

     > note: position the cell so light passes through the clear walls of the cell

   - press [0 ABS/100%T] to set the blank zero
   - remove blank; insert sample
   - measurement appears on the display

Glossary

buret a narrow graduated column with a stopcock at one end; a buret is typically used for adding a defined volume of a liquid to another solution. 27

colorimetric based on the development of color, usually assessed using a visible wavelength spectrophotometer. 183

K_a The dissociation constant for an acid, omitting the concentration of water as a constant. In aqueous solution, the K_a is equal to $[H_3O^+][A^-] / [HA]$. 19

molarity The number of moles of a solute per liter of solution. 19

osmolarity The molarity of particles in a solution – e.g., a 1M solution of a nondissociable solute is 1 Osmolar (the solution contains 6.023×10^{23} particles per liter). A 1M solution of a dissociable salt is n Osmolar, where n is the number of ions yielded from dissociation of one salt molecule.. 19

pH The negative logarithm of the molar concentration of protons in a solution (i.e., $pH = -\log[H^+]$). 19

pK_a The negative logarithm (base 10) of the K_a. 19

regression a statistically-derived formula describing an extrapolated curve through a given set of data points. 185

Bibliography

[1] M. Alley. *The craft of scientific writing.* Springer Verlag, 1996.

[2] P. Andrews. "Estimation of the molecular weights of proteins by Sephadex gel-filtration". In: *Biochemical Journal* 91 (May 1964), pp. 222–233.

[3] Author unknown. *Purification of salivary alpha amylase.* 2009. URL: http://www.science.smith.edu/departments/Biochem/Biochem_353/amylase.html.

[4] J. A. Beeley. *Fascinating families of proteins: electrophoresis of human saliva.* 1993.

[5] M. M. Bradford. "A rapid and sensitive method for the quantitation of microgram quantities of protein utilizing the principle of protein-dye binding". In: *Analytical Biochemistry* 72 (May 1976), pp. 248–254.

[6] M. Cargill and P. O'Connor. *Writing scientific research articles: strategy and steps.* Blackwell Pub, 2009.

[7] D. H. Chace et al. "Electrospray tandem mass spectrometry for analysis of acylcarnitines in dried postmortem blood specimens collected at autopsy from infants with unexplained cause of death". In: *Clinical Chemistry* 47 (2001), pp. 1166–1182.

[8] R. A. Day and B. Gastel. *How to write and publish a scientific paper.* 2006.

[9] X. Dong et al. "PlasMapper: a web server for drawing and auto-annotating plasmid maps". In: *Nucleic acids research* 32.suppl 2 (2004), W660–W664.

[10] R. Dybkaer. "The tortuous road to the adoption of katal for the expression of catalytic activity by the General Conference on Weights and Measures." In: *Clin Chem* 48.3 (Mar. 2002), pp. 586–590.

[11] B. Ganong. *Acid-Base Chemistry in Water*. 2008. URL: http://faculty.mansfield.edu/bganong/biochemistry/phofgly.htm.

[12] N. Gehlenborg et al. "Visualization of omics data for systems biology". In: *Nature methods* 7.3s (2010), S56.

[13] A. Glasfeld. *Intermolecular Interactions in Crystalline Glycyl-L-threonine Dihydrate*. 2006. URL: http://www.cgl.ucsf.edu/home/glasfeld/tutorial/GT/GT.html.

[14] B. Gustavii. *How to write and illustrate scientific papers*. Cambridge Univ Pr, 2008.

[15] M. N. Hegde. *A coursebook on scientific and professional writing for speech-language pathology*. Delmar Pub, 2009.

[16] X. Huang and W. Miller. "A time-efficient linear-space local similarity algorithm". In: *Advances in Applied Mathematics* 12.3 (1991), pp. 337–357.

[17] D. Lindsay. *Scientific Writing= Thinking in Words*. Csiro Publishing, 2011.

[18] M. Matsushita et al. "Determination of proteins by a reverse biuret method combined with the copper-bathocuproine chelate reaction". In: *Clinica Chimica Acta* 216 (July 1993), pp. 103–111.

[19] J. R. Matthews and R. W. Matthews. *SUCCESSFUL SCIENTIFIC WRITING: A STEP-BY-STEP GUIDE FOR THE BIOLOGICAL AND MEDICA*. Cambridge University Press, 2007.

[20] B. D. McCullough. "Fixing Statistical Errors". In: (2009). URL: http://www.csdassn.org/software_reports/gnumeric.pdf.

[21] B. D. McCullough and D. A. Heiser. "On the accuracy of statistical procedures in Microsoft Excel 2007". In: *Computational Statistics & Data Analysis* 52.10 (2008), pp. 4570–4578.

[22] J. N. Orvis and J. A. Orvis. "Throwing Paper Wads in the Chemistry Classroom: Really Active Student Learn-

ing." In: *Journal of College Science Teaching* 35.3 (2005), p. 3.

[23]   M. Schramm and A. Loyter. "Purification of a-amylases by precipitation of amylase-glycogen complexes". In: *Methods in Enzymology* VIII (1966).

[24]   M. Somogyi. "Modifications of two methods for the assay of amylase". In: *Clin Chem* 6 (Feb. 1960), pp. 23–35.

[25]   A. Tramontano, K. D. Janda, and R. A. Lerner. "Catalytic antibodies". In: *Science* 234 (Dec. 1986), pp. 1566–1570.

[26]   J. M. Word et al. "Visualizing and quantifying molecular goodness-of-fit: small-probe contact dots with explicit hydrogen atoms". In: *Journal of molecular biology* 285.4 (1999), pp. 1711–1733.

[27]   M. R. Wright. *An introduction to aqueous electrolyte solutions*. Wiley, 2007.

[28]   A. Yalta. "The accuracy of statistical distributions in Microsoft® Excel 2007". In: *Computational Statistics & Data Analysis* 52.10 (2008), pp. 4579–4586.

[29]   T. Zor and Z. Selinger. "Linearization of the Bradford protein assay increases its sensitivity: theoretical and experimental studies." In: *Anal Biochem* 236.2 (May 1996), pp. 302–308. URL: http://dx.doi.org/10.1006/abio.1996.0171.

CPSIA information can be obtained
at www.ICGtesting.com
Printed in the USA
LVHW060023220221
679574LV00006B/457